# Psychosocial factors at work and their relation to health

# PSYCHOSOCIAL FACTORS AT WORK

## AND THEIR RELATION TO HEALTH

Edited by

**Raija Kalimo**
**Mostafa A. El-Batawi**
**Cary L. Cooper**

**WORLD HEALTH ORGANIZATION**
**GENEVA**
**1987**

*The authors alone are responsible for the views expressed in this publication.*

ISBN 92 4 156102 5

TYPESET IN INDIA
PRINTED IN ENGLAND

85/6646 – Macmillan/Clays – 6000

# Contents

# Preface

Since 1974 the Member States of WHO have given increased attention to the question of psychosocial factors in relation to health and human development. The World Health Assembly asked the Director-General to organize multidisciplinary programmes that would explore the role of such factors and to prepare proposals for strengthening the Organization's activities in this field.[1]

The working population constitutes a major section of the community and industrialization in the developing countries and the automation of industrial processes in the more developed countries have given rise to rapid changes in the psychosocial environment at workplaces and in the reactions of the workers. Exposure to psychosocial stress at work is associated with a number of health problems, including behavioural disorders and psychosomatic disease.

Little attention has so far been paid, by those concerned with occupational health, to determining and controlling the psychosocial factors at work that lead to adverse health effects. The mental health problems of workers have continued to be regarded from the point of view of established psychiatric disorders requiring referral for treatment, and rehabilitation. In the field of ergonomics, efforts have been made to adapt machines and working processes to human physical and psychological capacities and limitations, but the guiding principles for day-to-day practice in the assessment of occupational psychosocial factors and their effect on workers' health have yet to be considered.

Those working in the occupational health services have therefore to face the fact that, while allowing for differences in the individual life-styles and susceptibilities of the workers themselves, psychosocial factors associated with new working methods are emerging as one of the significant causes of ill-health among working populations.

This book presents a review of the present state of knowledge. It is intended for both policy makers and health personnel, to enable them to establish appropriate occupational health services, develop methods for the monitoring and evaluation of the psychosocial working environment and the health status of exposed workers, and see where further research is needed.

---

[1] Resolution WHA27.53. *WHO handbook of resolutions and decisions of the World Health Assembly and the Executive Board*, Volume II, Geneva, World Health Organization, 1985, p. 101.

With advances in technological processes and in the control of physical and chemical hazards in the working environment it should be possible to reduce occupational disease. If this could be achieved, and if biological and psychosocial influences could also be controlled, there is no doubt that work in itself would more effectively fulfil its acknowledged role as a major factor in maintaining physical and mental wellbeing.

# Acknowledgements

The World Health Organization and the editors are pleased to acknowledge the guidance and support provided by Dr C. Husbumrer, Director, Division of Occupational Health, Ministry of Health, Bangkok, Thailand in planning the book, and wish to thank also Dr A. Jablensky, Division of Mental Health, World Health Organization, Geneva, Switzerland, for his help at all stages, and the following participants for their valuable contributions at consultations held during the last stages of preparation: Dr S. E. Asogwa, Department of Community Medicine, University of Nigeria; Mrs M. Daleva, Institute of Occupational Health, Sofia, Bulgaria; Dr L. Levi, WHO Psychosocial Centre, Laboratory for Clinical Stress Research, Stockholm, Sweden; Dr J. F. O'Hanlon, Department of Work and Organizational Psychology, University of Groningen, Netherlands; Dr M. Smith, Motivation and Stress Research Section, National Institute for Occupational Safety and Health, Atlanta, GA, USA; Dr M. Stilon de Piro, Occupational Safety and Health Branch, Department of Working Conditions and Environment, International Labour Office, Geneva, Switzerland.

Financial and technical assistance were provided during the various stages of preparation by the National Institute for Occupational Safety and Health, Atlanta, GA, USA.

**Part one**

# Introduction

Chapter 1

# Psychosocial factors and workers' health: an overview

Raija Kalimo[1]

Psychosocial factors are recognized to be critical in both the causation and the prevention of disease and in the promotion of health. This is so for the health sciences in general and for occupational health in particular, since psychosocial factors are among the most important of those that influence the total health of a working population.

## Psychosocial factors and the relatedness of work to health

Practitioners in occupational health have observed that working conditions not only cause specific occupational diseases, but may play a much wider role among the many determinants of a worker's health. From this basis the idea of the relatedness of work to health in a broad sense has gradually arisen. While an occupational disease is defined as a disease caused by certain well-defined factors in the working environment, a health impairment said to be work-related may result from multiple causation, the working environment having been one cause to a greater or lesser extent (2).

Careful consideration of the nature of health impairment said to be work-related has resulted in more attention being given to the psychosocial factors. Scientists investigating the human factors involved in maintaining health in the early 20th century were already concerned with certain psychological work-load parameters. Monotony, for example, aggravated by the newly invented methods of scientific management introduced during the period of industrialization had evoked attention and was the subject of experimental studies.

Since that time numerous epidemiological studies have demonstrated that health *is* related to psychosocial factors at work, and that their role in relation to both health status and the causation of disease is relatively wide in scope. Psychosocial factors can contribute to the causation and aggravation of a disease and affect the outcome of curative and rehabilitative measures. They can also be used as a means of promoting action for health at work. The conceptual aspects of health in relation to working environment are discussed in more detail in Chapter 2.

[1] Institute of Occupational Health, Helsinki, Finland.

## Reaction to stress: early symptoms of health impairment

Studies on preventive measures in the area of occupational health have led to an increased emphasis on the detection of early indicators of health impairment. Unspecific symptoms, including diffuse aches and pains, disturbed sleep, apprehension, anxiety, and mild forms of depression, are relatively common among working populations. Although they may be of diverse etiology, they are often indicators of chronic work-related stress. Perceived symptoms may be accompanied by objectively measurable changes in the autonomic nervous system and hormonal function. These dysfunctions, if chronic, may lead to health impairment and a clinically definable disease state. The various effects of occupational stress on psychological, behavioural, and physiological functions are discussed in Part two.

## The challenge of change and long-term stressors

Stress as perceived in working people is linked to a diversity of factors in the working environment and in the social setting. Great social changes are taking place in both industrialized and developing countries. While industrialized countries are currently experiencing rapid automation in relation to both production and services, developing countries are at a more fundamental stage of transition, from traditional ways of life to new types of production and methods of work that are often planned by foreign experts. Rapid urbanization is always accompanied by mechanization in industry. A concurrent problem is that different social institutions often seem to develop independently. Lack of coordination among social institutions connected with industrial development, housing, and transportation may lead to problems that are reflected in the health status of the workers. Although industrial development usually raises the standard of living and thus enables many of the basic necessities for a healthy life to be provided, it has its costs in that, sometimes, the resources for adaptation of the workers involved are exceeded.

Psychosocial problems in the field of occupational health cannot, however, be considered only on the basis of the changes currently taking place in working environment and occupational structure, or related social and familial phenomena. In many instances, a second or third generation of workers is involved— e.g., in mass production, using repetitive and monotonous routines, paced by machines, with little control over what happens at work. In many instances middle-class work ethics are deeply rooted, and professional work practices and organizational structures are firmly established. Still, a considerable number of employees in white-collar occupations are known to suffer from work pressures, such as those resulting from complexity, conflicts in roles and responsibilities, and the challenge of continuous competition.

Many of the stressors that people encounter in their daily working environment and through their social roles are discussed more extensively in Part three.

## Individuality: not a hindrance to constructive action

The relation between psychosocial factors at work and health is complicated by a large number of individual and subjective factors. An individual is rarely exposed to psychosocial influences from the working environment in isolation. Past experiences, genetic factors, and current general life conditions form a basis for an individual's own perception and interpretation of the influences from the working environment. Accordingly, reactions to perceived situations and the capacity to cope with, and recover from, periods of stress, are to a certain extent individually determined. This is discussed more extensively in Part four.

Individual differences must not, however, be overemphasized to the extent that group-oriented preventive intervention in working conditions is considered to be of secondary importance. When the influence of a psychosocial factor is strong, individual susceptibility is of lesser importance. An individual has many psychological characteristics and behavioural habits that are shared by others in a working population. This situation is analogous to that in which measures are adopted for controlling a physical disease; individuals may vary in their susceptibility and resistance to a disease for which a known pathogen has been determined, but the importance of fighting against the disease is not lessened because only a part of the population is affected.

## Work can be health promoting

From the health promotion point of view, work is invaluable. For example, it gives an individual a sense of belonging to a part of society considered to be important, needed, and valued; it provides an opportunity to express aptitudes and exercise, enhance, and acquire skills; it allows an individual to enter into the kind of social environment where a place can be found in goal-oriented interaction with others and mutual support in an interpersonal network; it provides opportunities to assume a variety of functions in carrying out daily activities, thus increasing possibilities for realizing different facets of the personality; and it creates a necessary time-frame structure. Work also usually brings the financial reward necessary to provide for the basic necessities. Thus, many of the essential ingredients of life satisfaction, health and wellbeing are inherently interwoven with work and occupation. The positive aspects of work should be awarded the merit they deserve and emphasized. In Part five approaches to promoting health in the working environment are dealt with, as well as the possible action to control known psychosocial health hazards.

## Values and knowledge: critical elements in setting priorities

A crucial factor to be considered in planning and implementing psychosocial intervention in the occupational setting is that the situations of working populations in various parts of the world vary considerably. In many countries the work force still has to strive to satisfy the needs fundamental to survival. In others the workers have some freedom to fulfil their own wishes for job variety and participation and have access to working situations that allow for physical, psychological, and social security. No matter what an individual's position is in the hierarchy of human needs and values, psychosocial considerations, in one form or another, must always be taken into consideration. Governments set their own priorities, however, and they may vary tremendously in quality and scope; health in the widest sense could be relatively low in the hierarchy of values influencing the production sphere. While it is understood that economic targets must be reached it should not be at the expense of the quality of life.

Inadequate attention to psychosocial issues may be due to insufficient knowledge. In the past few decades research activities in many countries have produced ample information on the interrelation of occupational factors and health. The task of those who have had experience in the systematic evaluation of the effect of work stressors on health is to disseminate information on that experience. Although only a very small amount of such information has spread beyond the narrow circle of professional expertise, interest has been raised and the demand is growing. An increasing flow of information through the mass media and widening cross-cultural exchange, making working populations aware of the importance of paying more comprehensive attention to the preconditions for health, will create expectations to which occupational health professionals should be ready to respond.

In recent research activities the important issue of material versus human values—i.e., the prerequisites for productivity and the health and wellbeing of the workers—has been brought to the fore and it has been recognized that they are not necessarily in contradiction. Dissatisfied, unmotivated, and stressed workers tend to have more health problems. Sick people tend to be less productive, to be absent from work more often, and to be more inclined to change jobs. Job satisfaction, health, organizational commitment, and productivity go hand in hand. This is recognized by many companies and employers who are pioneers in the area of occupational health, and who have established psychosocial programmes for organizational development and the humanization of work (1). The methods and the types of action employed in companies by experts in occupational health and by individual workers themselves are discussed in Part five.

## Comprehensive participation: essential for psychosocial improvements at work

It is an illusion to think that achievements relating to psychosocial factors at work depend on one-way communication from professional specialists to those who are active in the labour market. Continuous dialogue is the only approach to realistic and meaningful progress; though this is, regrettably, too often forgotten. The idea of workers' health can only be fully realized if the workers themselves participate actively and interact with others, such as managers, personnel administrators, and supervisors. Experts in occupational health have an important contribution to make in this regard. Recommendations on this type of coordination are given in Part five, and on research in Part six.

In spite of the fact that priorities differ from country to country and that, among various occupational groups, the levels of attainment of valued goals also differ, the number of those who would deny the overall importance of a psychosocial approach to the protection of workers' health is diminishing. It is hoped that national policy makers and legislators will institutionalize protective activities that are constructive and define who is responsible for carrying them out. Some govern-ments are already revising or formulating labour protection laws (see Chapter 16). Such legislative measures oblige those who are otherwise uninformed, or who are antagonistic, to strive to protect the health of the working populations with whom they are concerned.

## References

1 KANAWATY, G. ET AL. *Managing and developing new forms of work organization,* 2nd edition. Geneva, International Labour Office, 1982.

2 WHO Technical Report Series, No. 714, 1985 (*Identification and control of work-related diseases:* report of a WHO Expert Committee).

Chapter 2

# Definitions and the conceptual aspects of health in relation to work

Lennart Levi[1]

The evidence that *physical* stimuli in the occupational setting may cause physical disease—in the sense that exposure to them, or avoidance or manipulation of them, increases, decreases, or removes the chance of becoming ill, or reverses ill health when it occurs—is established for a large number of stimuli and diseases. The role of extrinsic *psychosocial* stimuli is not so clear. Before it is reviewed, therefore, some terms will be defined (2, 3, 4, 5, 8).

## Definitions

In the context of stress at work, *psychosocial stimuli* originate in a social process within a social structure and affect the organism through the mediation of perception and experience—the higher nervous processes—and may be suspected, under certain circumstances and in certain individuals, of causing disease. A factory, an enterprise, a school, a community, or a family are examples of a social structure. A social process is what is taking place in such a structure—e.g., work, education, care.

Psychosocial stimuli operate on man, and man is characterized by an individual *psychobiological programme*, a propensity to react in a certain pattern—e.g., when solving a problem or adapting to an environment. This propensity is, in turn, conditioned by genetic factors and earlier environmental influences.

The interaction between, or misfit of, environmental opportunities and demands, and individual needs, abilities, and expectations, elicit reactions. When the fit is bad, when needs are not being met, or when abilities are over- or undertaxed, the organism reacts with various *pathogenic mechanisms*. These are cognitive, emotional, behavioural, and/or physiological and, under some conditions of intensity, frequency, or duration, and in the presence or absence of certain interacting variables, they may lead to *precursors of disease*.

Examples of *cognitive* pathogenic mechanisms are restriction of the scope of perception (tunnel vision) or a lowered ability to concentrate, be creative, or make decisions. Examples of *emotional* pathogenic mechanisms are feelings of anxiety,

[1] WHO Psychosocial Centre, Laboratory for Clinical Stress Research, Karolinska Institute, Stockholm, Sweden.

depression, or alienation, mental fatigue, apathy, and hypochondriasis. Examples of *behavioural* pathogenic mechanisms are the abuse of alcohol, tobacco, or drugs, unnecessary risk-taking in working life and in traffic, and unprovoked aggressive and violent behaviour towards a fellow human being or towards oneself (suicidal behaviour). Some *physiological* pathogenic mechanisms are related to a specific situation, person, or disease. Others are nonspecific and these Selye termed *stress* (6). Technically speaking the term stress denotes a force that deforms bodies. Translated into everyday language this is more or less the same thing as load or pressure. In biology, however, the term stress often takes on a different meaning, being used to signify stereotype physiological "strain" reactions in the organism when it is exposed to various environmental stimuli—*stressors*—e.g., to changes in, or pressures and demands for adjustment from, the environment (4, 7). *Precursors of disease* are malfunctions in mental or physical systems that have not yet resulted in disability but that, if they continue, will do so.

*Health* is not merely an "absence of disease or infirmity" but also "a state of physical, mental, and social wellbeing" (10). *Wellbeing* is a dynamic state of mind characterized by reasonable harmony between a worker's abilities, needs, and expectations and environmental demands and opportunities.

The individual's subjective assessment is the only valid measurement of wellbeing available, even though it may not coincide with the objective views of others—for example, a worker may experience a sense of wellbeing while performing a monotonous or even potentially dangerous task.

Closely related to wellbeing is the concept of *quality of life*. By this is meant a composite measure of physical, mental, and social wellbeing (1, 9). When assessing the factors affecting wellbeing it must be recognized that the same factor may be good for some individuals but bad for others, or good in some situations and bad in others. Failure to recognize this and to consider the entire pattern of complex and non-linear interaction is probably one explanation for much of the confusion and controversy in this field.

Discussions of occupational stress often tend to omit *physical stimuli* in the working environment, in spite of the fact that they can influence the worker not only physically or chemically—e.g., a direct effect on the brain from an organic solvent—but also psychosocially. The psychosocial effects can be secondary to the distress caused by, say, odour, glare, noise, extremes of temperature and humidity, etc., and they can be due to the worker's awareness, suspicion, or fear that he is exposed to a life-threatening chemical hazard or to the risk of accident. Thus, exposure to an organic solvent can affect the human brain directly, whatever the worker's state of awareness, or his feelings and beliefs; it can also influence him indirectly, secondary to the unpleasant smell; and it can affect him because he knows or suspects that it may be harmful (5).

## The concept

The flow of events described above—social structure and process → psychosocial stimuli + psychobiological programme → pathogenic mechanisms → precursors of disease → disease—is modified by various *interacting variables* (Fig. 1). These are intrinsic or extrinsic factors, social, mental, or physical, that alter the action of causative factors at the mechanism, precursor, or disease stage—i.e., promote or prevent the process that might lead to disease. An example of an intrinsic preventive variable is coping (see Chapter 4) and an example of an extrinsic preventive variable is social support (see Chapter 15).

All of these interactions take place in a man–environment ecosystem. The process described above is not a one-way flow in a simple, linear, or even multifactorial model but constitutes a non-linear, cybernetic system with continuous feedback. Accordingly, if disease has occurred in an individual, it has repercussions on the social process within the social structure in

Fig. 1. A theoretical model for psychosocially mediated disease[a]

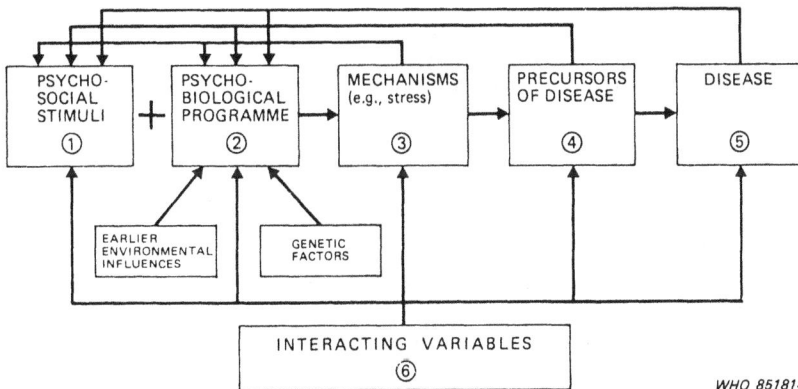

WHO 851814

[a] Reproduced by permission of the publishers from reference 3.

which he lives and on the resulting psychosocial stimuli (1), on his propensity to react (2), and on the interacting variables (6).

Although it is often possible to categorize factors according to the above definitions, there are occasions when the category is not clear or when categories are interchangeable. Nevertheless, it is hoped that to have provided them will facilitate discussion and lead to a better understanding of the problems.

## An example of the interaction of the components of the man–environment ecosystem

On the basis of economic and technical considerations, the management of a factory decides to divide the work process into very simple and primitive elements to be performed at an assembly line. By this decision a *social structure and process* is created, which constitutes the starting point of a sequence of events.

The situation is perceived by one of the workers involved and thus becomes a *psychosocial stimulus* (1). However, he has had extensive professional training and consequently hoped to be given a reasonably responsible work assignment. In addition, his past experience of work at the assembly line has been strongly negative—i.e., earlier environmental influences have conditioned his *psychobiological programme* (2). Further, hereditary factors make him prone to react with increases in sympathetic nervous activity and blood pressure—i.e., genetic determinants of the psychobiological programme (2). His wife blames him for the assignment and refuses to offer social support—an *interacting variable* (6). As a result, he feels depressed, increases his alcohol consumption, and his blood pressure rises—*mechanisms* (3).

The experience at work and in the family continues, and the worker's reactions, originally transient, become prolonged—*precursors of disease* (4). Eventually, a chronic depressive state and/or alcoholism and/or hypertension develop—*disease* (5). This, in turn, influences his environment at work and in the family and his psychobiological programme, possibly resulting in a vicious circle of interactions. Such a sequence of events can, of course, be counteracted or even prevented by modifying or improving the situation at work, transferring the worker to another job, decreasing his vulnerability, increasing his capacity to cope actively—to change the situation—or passively—to accept what cannot be changed—and encouraging fellow workers to offer social support.

## Socioeconomic and cultural setting

Health and wellbeing, or lack of them, therefore depend to a large degree on the characteristics, among them urban and rural environmental influences, of the socioeconomic and cultural setting in which the process is taking place. Environmental factors such as climate, geographical conditions, and the technology being used may also be decisive. Crowding, for example, in a small family enterprise with close friends and relatives creates very different reactions from crowding in a huge industrial plant with competitive strangers. Economic factors greatly modify individual and group reactions, because *inter alia* adequate economic resources make it possible to prevent or compensate for many potentially bad effects. Similarly, cultural factors strongly condition attitudes towards management, fellow workers, subordinates, and every other aspect of the working environment and the work itself. They also influence female and male roles and relationships at work and outside it, the age of entry into working life, and whether conditions are accepted passively as they are, or efforts to improve them are actively pursued.

## References

1 BESTUZHEV-LADA, I. V. & BLINOV, N. M., ed. *The modern conceptions of level of life, quality of life and way of life.* Moscow, USSR Academy of Sciences, Institute for Social Research, Soviet Sociological Association, 1978 (in Russian).

2 ELLIOTT, G. R. & EISDORFER, C., ed. *Stress and human health: analysis of implications of research.* New York, Springer, 1982.

3 KAGAN, A. R. & LEVI, L. Health and environment—psychosocial stimuli: a review,. In: Levi, L., ed. *Society, stress and disease: childhood and adolescence.* London, New York, and Toronto, Oxford University Press, 1975, Vol. 2, pp. 241–260.

4 LEVI, L., ed. Stress and distress in response to psychosocial stimuli. *Acta medica Scandinavica*, Suppl. 528, 1972, Vol. 191.

5 LEVI, L. *Preventing work stress.* Reading, MA, Addison-Wesley, 1981.

6 SELYE, H. A syndrome produced by diverse nocuous agents. *Nature*, **138**: 32 (1936).

7 SELYE, H. The evolution of the stress concept—stress and cardiovascular disease. In: Levi, L., ed. *Society, stress and disease: the psychosocial environment and psychosomatic diseases.* London, New York, and Toronto, Oxford University Press, 1971, Vol. 1, pp. 299 311.

8 SELYE, H. Preface. In: Selye, H., ed. *Selye's guide to stress research.* New York, Van Nostrand Reinhold, 1980, Vol. 1, pp. 8–13.

9 Szalai, A. & Andrews, F. M., ed. *The quality of life: comparative studies.* Beverly Hills, CA, Sage Publications, 1981.

10 World Health Organization. Constitution of the World Health Organization. In: *Basic documents,* Thirty-sixth edition, 1986, p. 1.

# Chapter 3

# The psychosocial health problems of workers in developing countries

Mostafa A. El-Batawi[1]

## Introduction

Many publications are available on the impact of industrialization on health. Studies have also been carried out on the health problems associated with urbanization. There have been very few investigations, however, on psychosocial factors at work and their health effects in developing countries. The following account is, therefore, essentially limited in scope.

Psychosocial factors may be classified in two main categories: those that have adverse effects on health, and those that may contribute positively to the worker's wellbeing. The problems of most concern in developing countries appear to be those associated with the need to adapt to the rapid changes in working and living conditions that result from industrialization and mechanization. In addition, there are the problems associated with the internal and external migration of workers in industry seeking to improve their earnings.

The following general remarks relate to the psychosocial environment in developing countries:

(1) The sociocultural environment is a determinant in the pattern of diseases and the frequency with which they are encountered.

(2) The socioeconomic status of a worker contributes to his susceptibility to psychosocial disorders. Furthermore, a worker's educational background considerably affects his reaction to industrial change and associated stress; in developing countries educational background among groups of workers varies widely.

(3) Religion and cultural pattern greatly influence human reaction to stress and thus any resulting health effect.

(4) Economic conditions play a role in shaping psychosocial reaction and the capacity to cope with occupational stress. High unemployment rates and under-employment may motivate the acceptance of normally stressful types of employment.

(5) Exposure to chemicals or adverse physical conditions in the working environment plays a role in shaping the psychosocial environment. Quite often, the existence of adverse working conditions leads to combined, and probably aggravated, effects on the worker's health.

[1] Chief Medical Officer, Occupational Health, World Health Organization, Geneva, Switzerland.

(6) Small factories and artisanal or family workshops usually present a more favourable psychosocial atmosphere than larger establishments. The predominance of small-scale industry in developing countries may, therefore, mean that there is a better overall psychosocial environment at work than in highly industrialized countries.

## Adverse psychosocial work factors and resulting effects

Workers in developing countries often have to make the transition from ' rural life, with its quiet and close relationships, to the factory environment; from traditional dependence on natural processes in agriculture and manual labour to standardized production, precise timing, rapid output, and dependence on energy; and from an identification with the land and crops to the impersonal environment of the machine. Such a transition requires an effort in adaptation that is usually compensated for by the material rewards of organized employment, which, in turn, provide the motivation to meet the challenges of adaptation. In many instances, however, adaptation fails, because of either too much stress imposed by the type of work, or personal susceptibility, or both; this results in absenteeism or in psychological and psychosomatic disturbance.

In some instances, such as in mechanized iron and steel production, the amount of work may be excessive or difficult for the worker to handle in relation to what he has been used to. This "overload" has also been associated with increased absenteeism and labour turnover (4). In other instances the work is repetitive and monotonous, of long duration without interruption, such as on assembly lines in electronic plants. In one country this, together with the susceptibilities of the workers and their failure to adapt, resulted in a general atmosphere of fear and in episodes of mass hysteria, leading to the closure of the factories affected (2, 6). In yet other instances, the work may be hazardous to health and life, as in deep-mining operations. In a developing Asian country, miners experienced considerable stress as they saw their colleagues killed in underground collapses, or suffering from pneumoconiosis due to exposure to dust. As a result a high labour turnover among the miners was observed (3).

In the early 1960s a relatively high incidence of peptic ulcer among workers at a glass factory was reported from a developing country. It was found that, in addition to diet and minor physical stress in the working environment, the main factor associated with what proved to be a significantly high incidence of acute peptic ulcer was the way in which the wages were managed in seeking increased output from individual glass blowers. They were paid by the piece, receiving more money per piece as the working shift progressed. This led to intense provocation and the workers were tired and irritable.

## Psychosocial factors in labour migrations

The health of migrant workers is a universal problem in both developing and developed countries and the psychosocial aspects of labour migration are of importance for this review because many developing countries are in a position to control any negative influences drawing on the experience gained from studies in developed countries at the early stages of their industrialization.

The migrant worker is generally of rural origin and is abruptly transplanted into an urban and foreign society; he would find it difficult enough to adapt to city life in his own country. The further away he is from his own country the greater are the difficulties. The migrant from a bordering country, with a similar culture, may adapt easily; this is not so with the migrant from another continent, who is not only ignorant of the environment in which he is to live but has to contend with the considerable differences between that environment and the one from which he has come. His cultural background, customs, and traditions often create barriers to integration within the host country. Everything is different: climate, eating habits, social customs, cost of living, housing conditions, and type and rhythm of work. The migrant is handicapped by his inexperience in regard to urban life, his inadequate knowledge or complete ignorance of the language, his illiteracy, and probably by his lack of occupational skills. He comes to realize that any skills and experience he does have count for very little if they have no relevance to the requirements of his new activities. He is, therefore, trapped from the beginning in the sort of contradiction that can have pathological consequences. This is compounded by emotional factors; fear of isolation, loneliness, sadness at separation from his family, and fear of losing the job for unforeseen managerial or economic reasons. Such factors greatly influence his behaviour and can predispose him to illness.

The prevalence of psychiatric disorders appears to be 2–3 times as high among recent migrants as among the local population. The etiology of these disorders lies in the de-personalization phenomenon in those migrant workers who are unable to adapt themselves to the surrounding cultural environment. They may manifest themselves as various somatic conditions, particularly of the digestive tract (5). The somatic and psychological disorders attributed to migration include duodenal ulcer, reactive depression, personality disturbances, and, in extreme cases, psychotic illness (1).

## Occupational psychosocial factors and the wellbeing of workers

The Director-General's progress report on the WHO occupational health programme presented to the Thirty-second World Health Assembly stated that "Work, the key element to progress and achievement, is the human being's main identification with a productive life. It is a human objective as well as a means of earning a living. The continuous interaction between man and his physical and psychological working environment may influence his health either positively or negatively, and the production process itself is influenced by the worker's state of physical and mental wellbeing. Work, when it is a well-adjusted and productive activity, can be an important factor in health promotion—an aspect that has not as yet been exploited to the advantage of the nations' health" (8).

Thus, the positive aspects of work and its effect on health should not be underestimated. Many employers in developing countries are maintaining long-inherited humane traditions through close and friendly working relations with their employees, particularly in the small establishments that employ the bulk of the productive manpower (4). Despite the many undesirable physical hazards in small factories, there is evidence that the rate of absenteeism is lower than in larger establishments, which may be attributable to the satisfactory working relations in the former. In larger establishments in many developing countries, religious traditions are upheld; a church, temple, or mosque is built on the factory premises and the workers are allowed to pray there.

The positive psychosocial effects of working can be demonstrated by looking at the health consequences of unemployment, which include depression, anxiety, and increased morbidity (7).

## Means of control

Much of the action necessary to prevent psychosocial problems in workers is of a social, political, or economic nature and, therefore, extends beyond the scope of the occupational health services alone. However, the occupational health services can make a major contribution in the form of advice, information, and recommendations, and activities in cooperation with other health and social disciplines. They remain in the front line for the primary prevention of the adverse effects of psychosocial factors at the work place. Among the responsibilities of the staff of the occupational health services, are the following:

— determination of stress factors in the working environment and among workers;
— recognition of changes in behaviour at an early stage;
— initiation of action to prevent ill-responses;
— coordination with other services, management, and labour unions with a view to prevention.

Most preventive measures can be initiated by nurses, social workers, or primary health care workers and the workers themselves should participate as a means of receiving guidance for the future. Preventive measures should include:

(1) A *pre-placement examination* aimed at assessing the worker's psychological tendency; it should include a review of his psychological history. An aptitude test and information on his job expectations and capacity for coping with demanding tasks would help in proper placement.

(2) A *periodic examination* during which the occupational health physician and his staff look for any significant behavioural change and/or psychological or psychosomatic disorder. Evidence of such a change should lead them to investigate the worker's personal history and the possible causative factors at work and elsewhere. The most significant types of change in behaviour appear to be frequent complaints of fatigue or of suffering from an ailment that has no organic basis or only to a very minor extent. This may often be the only way in which the worker can communicate his personal psychological problems. Even if it constitutes an increasing demand on the health and social services, special attention should be given to such complaints if they are not to develop into major psychogenic or behavioural disorders.

(3) A programme for *monitoring* the working environment and the health of the workers for psychosocial factors. There is a

need to develop the appropriate methodology for undertaking this essential type of monitoring.

(4) Visits to the workers' homes, families, and recreation places in order to obtain useful information on the general environmental and family conditions in which they live.

## Recommendations for field studies

There is an urgent need to initiate systematic studies of occupational psychosocial factors in developing countries. In addition:

(*a*) A means of monitoring the occupational psychosocial environment and the health of the workers must be found;

(*b*) Factors that may lead to excessive stress must be recognized and controlled at an early stage of industrial development, drawing on the experience of developed countries;

(*c*) The psychosocial factors that contribute to health must be determined and employers, the workers themselves, and health practitioners must be educated so that they can utilize such factors for the betterment of the workers' health;

(*d*) Cross-sectional studies and surveys, utilizing simple, practical means of investigation, should be encouraged.

## References

1 ALMEIDA, Z. *Aspects psychosociaux et psychopathologiques de la transplantation; la santé des migrants.* Paris, Edition Droit et Liberté, 1972, pp. 105–128.

2 CHEW, P. K. How to handle hysterical factory workers. *Journal of occupational health and safety,* **47**: 50–53 (1978).

3 CHO, K. S. & LEE, S. H. Occupational health hazards of mine workers. *Bulletin of the World Health Organization,* **56**: 205–218 (1978).

4 EL BATAWI, M. A. Psychosocial stressors in working life: problems specific to developing countries. In: Levi, L., ed. *Society, stress and disease: working life.* Oxford, New York, and Toronto, Oxford University Press, Vol. 4, 1981, pp. 12–13.

5 INTERNATIONAL LABOUR OFFICE. *Occupational health and safety of migrant workers:* Seventh report of the Joint ILO/WHO Committee on Occupational Health. Geneva, 1977 (Occupational Safety and Health Series, No. 34).

6 PHOON, W. O. Industrialization—at a price. *World Health,* November 1981, pp. 26–29.

7 TIFFANY, D.W. ET AL. *The unemployed: a social psychological portrait.* Englewood Cliffs, NJ, Prentice-Hall, 1970.

8 WORLD HEALTH ORGANIZATION. *Programme budget for the financial period 1980–1982: progress report by the Director-General—Occupational health programme.* Geneva, 1979, p. 5 (document A32/WP/1).

# Part two

# Reactions to stress

Chapter 4
# Psychological and behavioural responses to stress at work

Raija Kalimo[1] and Theo Mejman[2]

The psychological and behavioural manifestations of stress may take different forms and be of varying intensity. Sometimes there are no outward manifestations but those in distress suffer internally. At other times clearly observable, even dramatic, emotional and behavioural expressions of distress become apparent. To understand such reactions three crucial steps in the gradual development of job stress must be recognized, namely, perceived threat, manifestation of coping, and occasional failure to cope.

In this chapter are introduced the various means of coping that people adopt when under stress. The main emphasis is on the psychological and behavioural indicators of unsuccessful coping as a result of excessive situational demands, or limitations on an individual's own resources. It is not intended to make a comprehensive review of the subject but rather to discuss crucial issues from the point of view of occupational health, and to present a few examples based on empirical findings. Reactions to stress that relate primarily to the individual are discussed first, followed by the behavioural manifestations of stress that have direct relevance for the work of the organization that employs him.

## Coping with stress

The main causes of stress at work are the inadequate demands of a job in relation to the worker's abilities, frustrated aspirations, and dissatisfaction with regard to valued goals. Man is able to deal with these situations by means of a number of coping strategies. Coping has been defined as "efforts, both action-oriented and intrapsychic, to manage (i.e., master, tolerate, reduce, minimize) environmental and internal demands and conflicts which tax or exceed a person's resources" (22).

In certain situations a worker may try to cope by making an effort to change the situation for the better; at other times it may be possible to avoid an intolerable situation. If it is not possible either to change or to avoid the situation, for external or internal reasons, he may rely on a palliative mode of coping. A palliative mode of coping is a way of responding that helps a

[1] Institute of Occupational Health, Helsinki, Finland.
[2] Department of Work and Organizational Psychology, University of Groningen, Groningen, Netherlands.

person to feel better, in spite of the fact that the actual problem is not resolved. Under certain conditions and in the short-term such defensive reactions as denial, intellectualization, or repression of thoughts may prove to be adequate coping strategies. It has been said that the protective functions of coping are made real "by eliminating or modifying conditions giving rise to problems; by perceptually controlling the meaning of experience in a manner that neutralizes its problematic character; and by keeping the emotional consequences of problems within manageable bounds" (31).

Some coping strategies may be helpful only if applied in the short-term. For example, avoidance or escape by drinking alcohol may be a consciously controlled and mitigating way of coping occasionally. However, such behaviour could become, in the course of time, a less and less deliberate and increasingly desperate and forceful habit, with debilitating social and health consequences. There are coping strategies that may be appropriate in certain situations but not in others. Flexibility in adopting an appropriate coping strategy is one of man's main health resources. The capacity to deal with a stressful encounter can create a feeling of mastery and self-confidence and a short-term stressful experience can produce a gratifying sense of achievement and satisfaction.

## When coping fails

Psychosocial stressors at work or as a consequence of employment conditions are frequently long-standing, continuous, or often-repeated. In spite of the many ways in which a person can draw on his own resources to cope, the demands may exceed the resources, and his manner of coping may be inefficient, or in the long run a new source of problems. The results can be seen as disturbances in the psychological and behavioural functions. Among the early indicators are negative feelings, such as irritation, worry, tension, and depression. Cognitive disturbances, seen in a lowered performance capacity, may follow. Avoidance, originally aimed at coping and mastery, may turn into fixed, non-purposeful, obsessive behavioural disorder.

### Psychological symptoms

Among the most commonly measured long-term manifestations of stress are self-reported psychosomatic complaints, psychiatric symptoms, or complaints about wellbeing. Data, published in 1974, were reanalysed in 1977 in order to derive information on the association of perceived work factors and psychometrically measured stress effects (24, 25). The work factors included working time, work-load, working conditions, job content, non-participation in deciding how the job should be done, job uncertainty, and social isolation. The stress effects included psychosomatic complaints, general dissatisfaction with life, loss of self-esteem, and depression. The factor "working time" did not correlate with any effect. The highest rate of correlation was between loss of self-esteem and general job content—defined as possibilities for using and/or developing personal skills in the job—and variability of job content. Loss of self-esteem was also related to social isolation. The remaining correlation rates were low. Given that judgements on both the work factors and the stress effect measures were partially subjective, it was concluded that the correlations were generally weak. A similar conclusion can be drawn from an epidemiological study of samples representing 23 occupations in the USA (8). Loss of interest in work, alienation, and even a decrease in intellectual capacity have been reported as the outcome of a chronic lack of control over one's working situation (14, 20). The problem of monotony and/or understimulation is of great importance in industrial mass production and is becoming of concern in modern system monitoring and control. An extensive review of the consequences of boredom concludes that its expression in workers with monotonous occupations is associated with measurable attentional, perceptual, cognitive, and motor impairment that can substantially degrade performance efficiency (30). Although only a minority of workers experience severe, chronic boredom in even the most monotonous of occupations (estimates rarely exceed 30%) the degree of boredom experienced is a strong determinant in measuring job satisfaction. The same review of the consequences of boredom provides evidence from epidemiological studies of a relationship between monotony and ill health; and also of the fact that the degree of repetition in work is a major determining factor in the amount of absenteeism. Persons whose jobs are restricted and monotonous are less likely than those who hold more interesting jobs to engage in leisure activities requiring planning, participation, and effort. It is interesting to note from laboratory studies that understimulation has a facilitating effect on adrenaline production comparable to that of overstimulation (12). From the psychological point of view some differentiation has been observed:

qualitative overstimulation, or overload, is associated with tension and low self-esteem, whereas qualitative understimulation is associated with depression, irritation, and psychosomatic complaints (42). Both overstimulation and understimulation are equally associated with job dissatisfaction.

Comparisons have been made between groups of workers engaged in repetitive machine-paced tasks, repetitive self-paced tasks, and non-repetitive work (7). The stress indications were measures of anxiety, somatic symptoms, depression, and dissatisfaction. In all the groups moderate but consistent positive correlations were found between dissatisfaction and somatic symptoms. Those engaged in repetitive tasks reported more dissatisfaction than those engaged in non-repetitive work. However, repetitiveness as such was not associated with anxiety. Both machine- and self-pacing, on the other hand, were associated with anxiety and, to a lesser extent, with somatic symptoms and depression, though not with dissatisfaction. A similar conclusion was reached from a review of the data from the epidemiological study of samples representing 23 occupations in the USA (8, 19). Machine-paced assemblers were not more dissatisfied with their work than either self-paced assemblers or blue-collar workers in general. They differed, however, in three health measures: somatic complaints, anxiety, and depression. This difference is important as evidence of personal gratification and development. It also means that job dissatisfaction and mental health impairment are independent. The former is less specific and the latter more specific in relation to particular causal factors—e.g., machine-pacing. In any case, these two psychological effects of stress are not equivalent and as concepts should not be used interchangeably.

Recent reviews provide further evidence of the association of job-stressors with self-reported psychiatric symptoms and with level of self-esteem, feelings of wellbeing, boredom, resentment, and fatigue (4, 18, 19, 38, 42). The literature does provide evidence of relationships between occupational stressors and psychological symptoms but definite causality remains to be confirmed.

## Burn out

The expression *"burned out"* has become popular for describing a condition experienced by employees in professions involving a high degree of contact with other people. On the basis of a comprehensive review of studies on the "burn out" syndrome it was recently defined as "a response to chronic emotional stress with three components: (a) emotional and/or physical exhaustion; (b) lowered job productivity; and (c) overdepersonalization" (32). In reports on clinical experience and case analyses other symptoms for inclusion in a description of the burn out syndrome have been put forward, such as low morale, a negative attitude towards patients, clients, or similar types of person at work, a cynical attitude towards the achievement of working goals, exaggerated confidence expressed in overt behaviour, absenteeism, frequent changes of job, and other escapist behaviour such as using drugs.

The current evidence indicates that burn out must be regarded as multidimensional and cannot be expressed in terms of a single index. Further research is needed to analyse the structure of the syndrome. Its three components can be classified into the three major symptom categories of stress: physical symptoms—e.g., physical exhaustion; symptoms connected with attitudes and feelings—e.g., emotional exhaustion, overdepersonalization; and behavioural symptoms—e.g., overdepersonalization, lowered job accomplishment, lowered productivity. Thus, the response categories are probably the same as in any occupational stress situation, while the pattern may, to a certain extent, be specific to the type of work in which a person comes into contact with many others.

## Smoking and eating habits

Smoking is a habit that may have a number of both internal and external motives. It has been shown to be associated with tension and anxiety (26). A relation has been demonstrated between stress at work and smoking; the decision to stop smoking, in particular, has been shown to be negatively related to various job stressors (36, 39). In a study published in 1975 no differences were found among smokers, ex-smokers, and non-smokers in the USA, in relation to the stress they reported in connection with their jobs (8). However, an inability to stop smoking in engineers and scientists was associated with job stress and high levels of quantitative work-load.

Changed eating habits, overeating in particular, are reaction patterns observed relatively often during periods of intensive stress, but the data available are unsystematic. There is obvious justification for research into the contributing role of occupational psychosocial factors in obesity, while its control is of considerable public health importance.

## Use of alcohol

Increased or excessive alcohol consumption and escapist drinking behaviour are generally regarded as possible reactions to psychosocial problems at work. Empirical data to confirm that hypothesis are, however, scarce, one of the main reasons being the difficulty in obtaining reliable figures on alcohol consumption. In addition, this type of behaviour is regulated by cultural heritage and social norms and it is not possible to clarify its role uniformly as a stress response.

In a study carried out in the Stockholm area to investigate the reasons for alcohol consumption among working people, a questionnaire was sent to key informants at work: company managers and chiefs, physicians, and the representatives of groups of workers and labour unions (21). Work-related factors such as job dissatisfaction and time pressure belonged, with marital problems, to the most important reasons for drinking reported by the workers' representatives. The foremen regarded time pressure as the third reason in order of rank for alcohol consumption. The physicians placed work-related reasons, such as time pressure and feelings of insecurity, second in order of rank, following marital problems, which was regarded as the primary reason by all respondents.

A follow-up study was carried out in Scotland from 1975 to 1978 to see if professions where there is a greater probability of acquiring alcohol-related problems attract people already manifesting such problems, or heavy drinkers (33). New employees at selected breweries were first interviewed during their first three months in the job, with controls selected from an organization with a known low alcohol consumption rate. The new employees at the breweries had poorer vocational backgrounds and their alcohol consumption rates were greater than the workers in the control group. Their consumption of alcohol had increased at the new workplace and they drank more on the job than the controls. At the follow-up interviews 2–3 years later, while the alcohol consumption rate and the alcohol-related problems had increased in the breweries, there were no changes in the control group.

In a study published in 1974 workers in various occupations were asked about the presence of 6 given stressors at work and about various behavioural habits (24). Escapist drinking was related to both too little and too much work, the inappropriate use of knowledge and skills, insecurity of job tenure, and infrequent opportunity for participation in deciding how the job should be done. The estimated overall perceived work-load correlated with the rate of escapist drinking. Although these results point to many problems that theoretically could be reasons for escapist behaviour, including drinking, it is impossible to infer causal relations on the basis of the data obtained.

Among female auxiliary nurses in Finland the correlation between working conditions and alcohol consumption was low (17). Internal professional controversy, conflict between personal values and work, job dissatisfaction, unfulfilled expectations concerning work, and a lowering in commitment and sensitivity were related to alcohol consumption. Alcohol consumption correlated with perceived health status and was significantly related to perceived occupational stress among seamen (11).

Investigations in the police force in the United Kingdom indicated that high levels of work-related stress can facilitate the tendency of some individuals to drink heavily as a way of coping (10).

The medical records of ambulatory clients of an alcohol clinic in the USA were analysed during a six-year period (41). The alcoholics were characterized by low job satisfaction. More than half had problems in accepting their supervisors. Only about 20% felt they were working towards their personal professional objectives. About 25% had no occupational goals.

Employees in companies in the USA, referred for help because of alcohol-related problems, were compared with a matched group of employees without such problems (2). The data were collected during the first two weeks of visits by the problem drinkers (as ambulatory patients). Job satisfaction was about the same in both groups. Significant differences between the groups were found in their feelings regarding the attainment of life goals. Almost 50% of the alcoholics were disappointed with their achievements but only 21% among the control group.

The characteristics of work and working environment typical in the personal history of problem drinkers have been summarized as follows (35):

— *Lack of visibility.* Jobs that have nebulous production goals; that allow individual workers to exercise their own opinions in regard to schedules and output; that are remote from the observance of supervisors and associates.
— *Absence of structured work.* Jobs that allow the emergence of a stress factor such as addiction to work; that do not give the worker a specific role to play, causing him to be

ignored, or require him to perform tasks that are obsolete; that are newly created in the organization with what the worker is supposed to do as yet unclear.

— *Absence of social control.* Jobs in which the incumbent is required to drink in relation to his work; that are stressful and lack social controls, if they are occupied by incumbents who have moved from highly controlled jobs in which heavy drinking was practised to relieve tension.

## Work performance

One of the consequences of occupational stress which concerns organizations the most is variation in an employee's performance. Performance is directly connected with an organization's effectiveness and is thus linked with its economic interests more clearly than the other consequences of stress. Work performance has been widely studied as a function of various factors in the working environment but less often in connection with stress as a mediating factor (4). It has not been considered of great interest in studying the human consequences of occupational stress.

Eight performance variables potentially useful in the analysis of occupational stress have been differentiated (38). Among these are: adequacy of performance (the ratio of actual output to a determined quota); quality and quantity of output; erroneous responses, "misses", "blocks" (failures to initiate a response); and variability in work-cycle time. Caution is needed in the interpretation of performance. The performance of a worker is affected by loading or stressing factors but often indirectly. Other important factors that cannot be conceived as stressors, such as working strategy, influence the quality and/or quantity of performance. The relationship between stress and performance is non-linear (1). A model describing performance efficiency as an inverted U, i.e., positive quadratic, function of stress is the one most widely accepted (43). This means that a person performs at optimum level when subjected to moderate degrees of stress and less efficiently when the degree is either very high or low.

When a person has an inappropriate work-load and is under stress, his behaviour may change; for example, side-issues may be neglected to concentrate on the main task. Changes in the manner of carrying out a task, or strategy changes under conditions of overload, have been observed frequently in field studies and in the laboratory. Experienced workers change strategies to prevent overload and to reduce stress. The effectiveness of a strategy depends on the possibilities for implementing it in the operational environment and the capability of the worker himself (3). When faced with increasing numbers of aircraft to handle, experienced air traffic controllers chose to simplify their strategies (40). It was further demonstrated that experience determines the extent to which strategies are changed under such circumstances: the less experienced air traffic controllers persisted in employing the more precise but time-consuming strategies (6).

In an investigation involving industrial weavers it was found that productivity as well as the degree of stress are affected by the type of strategy selected (15). High and low productivity weavers were selected for study. The more productive used more efficient strategies than the less productive. The common strategy of the former was to anticipate and prevent breakdowns in the weaving process, so that they spent less time during the day repairing and correcting. The weavers in both groups were striving to achieve the same quotas, so the work-load of the less productive increased at the end of the day. Their greater level of stress was revealed by changes in several measures of behaviour and performance. The less productive weavers were taught to use anticipatory strategies, which improved their performance and subjected them to less stress.

## Absenteeism and turnover

Absenteeism due to sickness has increased in all the developed countries over the past few decades. The frequency of absences during the year has increased more rapidly than the number of lost working days, which indicates that the number of short periods of absence has increased more than the number of long periods.

Data regarding absences from work can be classified according to several criteria: reason for absence, duration of absence, and necessity for absence. The need to be absent because of a severe illness or an accident cannot be disputed. Absence due to mild illness, fatigue, job strain, or for a private reason is, however, to a large extent a matter of choice, depending on many factors both within and outside the working place, such as local and cultural permissibility, personal considerations in regard to loss of wages, and control policies exercised by the employer and the insurance carrier.

In a survey of 184 122 enlisted men in the United States Navy over a period of 11 years a relationship was shown between the

job-stress scores (ratings given by the researcher) and hospitalization rates for 10 stress-related illnesses, including alcoholism,
neuroses, hypertension, ischaemic heart diesase, and ulcers (16).

Repeated evidence indicates that absenteeism and turnover
are related to job dissatisfaction. A review of studies published
before 1973 draws the conclusion that job dissatisfaction is a
central factor in withdrawal (34). Level of job satisfaction, in
turn, is determined by a multitude of work-related factors. It is
thus necessary to investigate how the working environment can
affect absenteeism, information of basic importance for the
planning of preventive action.

In a number of studies specific attention has been given to the
work-related parameters in absenteeism and turnover, both of
which seem to be determined to a large extent by the same types
of work factor. In one study, in which a group of sawmill
workers at high risk to stress was compared with two control
groups consisting of individuals of the same age, 29% of the
sawmill workers had been absent from work for 30 days or more
during the preceding year (13). According to the criteria applied
43% of those absences were caused by stress. No member of
either control group had such a high rate of absence. Turnover
was related to role ambiguity in a study published in 1971 (23).
In a study of nurse's aides in the USA the level of perceived
occupational stress was related to the decision to leave the job
but it is not known whether the stress was the reason for
leaving (27). In the study published in 1974, in which workers in
various occupations were asked about the presence of 6 given
stressors at work and about various behavioural habits, positive
relationships were found between perceived job stressors—
resource inadequacy and too much work—and the frequency
with which workers complained formally to the management.
Three job stressors—the underutilization of skills, insecurity of
tenure, and non-participation in deciding how the job was
done—were inversely related to the workers' complaints (24).

Reviews of the literature (9, 28, 34) indicate that absenteeism
and turnover are related to the following factors in the working
environment:

— unmet expectations in regard to pay and incentives;
— few opportunities for promotion;
— lack of recognition from the supervisor in terms of comment
   and guidance on performance and discussion on an equal
   footing;
— an inexperienced supervisor;

— dissatisfaction with relationships with other workers;
— task repetitiveness;
— lack of responsibility and autonomy;
— work role ambiguity;
— large number of workers in one unit.

Interestingly, there seems to be a progression from a tendency to be late to absenteeism (9) and from absenteeism to changing jobs (28).

A study published in 1982 showed that there was a drastic reduction in absenteeism among employees who had been often absent after they had participated in a health evaluation programme aimed at helping them to cope with job stressors more effectively (37).

Sociodemographic factors are related to absenteeism to a relatively large extent. Young workers are more often absent than older ones. With increasing age, short absences tend to diminish and long absences to increase (5). The number of children the workers have and the day-care facilities available are additional determining factors, especially among women (29).

## Conclusions

Stressful experiences at work may manifest themselves in a number of psychological and behavioural reactions. A person normally copes with transitional periods of stress at work by either altering the situation or controlling his response. Many periods of stress, therefore, pass without noticeable reaction. Problems arise when working conditions are in opposition to human needs and resources over a long period of time, with a failure to cope in consequence. Negative emotions, tension, worry, and depression are some of the first signals of such a stressful situation, and may be accompanied by impaired cognitive functions and performance capacity. Further behavioural change, such as avoidance of, or escape from, the situation, either physically or mentally, may follow. Many reactions of this nature are evident only to the closest members of the sufferer's family. That is why work-related problems often develop into major psychological, behavioural, or physical disorders with consequent problems in relation to the fulfilment of occupational and other social functions.

Occupational health professionals are in key positions to recognize the psychological and behavioural indicators of stress at work at an early stage. They need to act to sensitize both workers and the general public to the problems that can develop, and to the fact that they can often be avoided if appropriate action is taken in time.

## References

1 ALLUISI, E. A. & FLEISHMAN, E. A., ed. *Stress and performance effectiveness.* Hillsdale, NJ, Erlbaum, 1982 (Human Performance and Productivity Series, Vol. 3).

2 ARCHER, J. Social stability, work force behavior, and job satisfaction of alcoholic and nonalcoholic blue-collar workers. In: Schramm, C. J., ed. *Alcoholism and its treatment in industry.* Baltimore and London, The Johns Hopkins University Press, 1977, pp. 156–176.

3 BAINBRIDGE, L. Problems in the assessment of mental load. *Le travail humain,* **37**: 279–302 (1974).

4 BEEHR, T. & NEWMAN, J. E. Job stress, employee health, and organizational effectiveness: a facet analysis model and literature review. *Psychological perspectives,* **31**: 665–699 (1978)

5 BEHREND, H. & POCOCK, S. Absence and the individual: a six-year study in one organization. *International labour review,* **114**: 311–327 (1976).

6 BOUJU, F. & SPERANDIO, J. C. *Effets du niveau de qualification et de la charge de travail des contrôleurs d'approche sur leur strategies opératoires.* Institut de Recherches Sociologiques Appliquées, Rocquencourt, 1978.

7 BROADBENT, D. E. & GATH, D. Chronic effects of repetitive and nonrepetitive work. In: Mackay, C. J. & Cox, T., ed. *Response to stress, occupational aspects.* Guildford, Surrey, IPC Science and Technology Press, 1979, pp. 120–128.

8 CAPLAN, R. D. ET AL. *Job demands and worker health: main effects and occupational differences.* Washington, DC, United States Government Printing Office, 1975 (DEHW Publication No. (NIOSH) 75-160).

9 CLEGG, C. W. Psychology of employee lateness, absence, and turnover: a methodological critique and an empirical study. *Journal of applied psychology,* **68**: 88–101 (1983).

10 DAVIDSON, M. J. & VENO, A. Stress and the policeman. In: Cooper, C. L. & Marshall, J., ed. *White-collar and professional stress.* London, Wiley, 1980, pp. 131–166.

11 ELO, A. L. *Merenkulkijoiden työ ja teaveys* [*Seafarers' work and health.*] Helsinki, Institute of Occupational Health, 1979 (Report No. 155) (in Finnish with a summary in English).

12 FRANKENHAEUSER, M. Psychoneuroendocrine approaches to the study of emotion as related to stress and coping. In: Howe, H. E. & Dienstbier, R. A., ed. *Nebraska symposium on motivation, 1978: Human emotion.* Lincoln, NE, University of Nebraska Press, 1979, pp. 120–161. (Nebraska Symposia on Motivation Series, Vol. 26).

13 GARDELL, B. *Arbeitsgestaltung, intrinsische Arbeitszufriedenheit und Gesundheit.* [*Work organization, intrinsic job satisfaction and health.*] In: Frese, M. et al., ed. *Industrielle psychopathologie.* Bern, Huber, 1978, pp. 52–111 (in German).

14 Greif, S. Intelligenzabbau und Dequalifizierung durch Industriearbeit? [Does work in industry degrade the intelligence and undermine skills?] In: Frese, M. et al., ed. *Industrielle psychopathologie*. Bern, Huber, 1978, pp. 232–256 (in German).

15 Hacker, W. & Vaic, H. Psychologische Analyse Interindividueller leistungsdifferenzen als ein Grundlage von Rationalisierungsbeiträge. [Psychological analysis of individual differences in performance as a basis for the rationalization of work.] In: Hacker, W. et al., ed. *Psychologische Arbeitsuntersuchung*. Berlin, DDR, Deutscher Verlag der Wissenschaften, 1973, pp. 109–131 (in German).

16 Hoiberg, A. Occupational stress and illness incidence. *Journal of occupational medicine*, **24**: 445–451 (1982)

17 Jokinen, M. & Pöyhönen, T. *Apuhoitajan työn stressi ja muut työsuojeluongelmat.* [*Stress and other occupational health problems afflicting practical nurses.*] Helsinki, Institute of Occupational Health, 1980 (Report No. 166) (in Finnish with a summary in English).

18 Kalimo, R. Stress in work: conceptual analysis and study on prison personnel. *Scandinavian journal of work, environment and health*, **6**: Suppl. 3, 1980.

19 Kasl, S. V. Epidemiological contributions to the study of work stress. In: Cooper, C. L. & Payne, R., ed. *Stress at work*. Chichester, New York, Brisbane, and Toronto, Wiley, 1978, pp. 3–48.

20 Kohn, M. L. & Schooler, C. *The reciprocal effects of the substantive complexity of work and intellectual flexibility: a longitudinal assessment*. Bethesda, MD, National Institute of Mental Health, 1977.

21 Kühlhorn, E. Spriten och jobbet. [Alcohol and work]. *Alkoholfraagen*, **65**: 222–230 (1971) (in Swedish).

22 Lazarus, R. & Launier, R. Stress-related transactions between person and environment. In: Pervin, L. A. & Lewis, M., ed. *Perspectives in international psychology*. New York, Plenum Press, 1978, pp. 287–327.

23 Lyons, T. F. Role clarity, need for clarity, satisfaction, tension and withdrawal. *Organizational behavior and human performance*, **6**: 99–110 (1971).

24 Margolis, B. K. & Kroes, W. H. Occupational stress and strain. In: McLean, A. ed. *Occupational stress*. Springfield, IL, Thomas, 1974, pp. 15–20.

25 Marstedt, G. & Scharn, K. Eine Analyse des Zusammenhang von Arbeitsbedingungen und Psychischen Störungen. [An analysis of the connections between working conditions and mental disorders.] *Psychologie und praxis*, **22**: 1–12 (1977) (in German).

26 McCrae, R. R. et al. Anxiety, extraversion and smoking. *British journal of social and clinical psychology*, **17**: 269–273 (1978).

27 McKenna, J. F. et al. Occupational stress as a predictor in the turnover decision. *Journal of human stress*, **7**(4): 12–17 (1981).

28 Muchinsky, P. M. Employee absenteeism: a review of the literature. *Journal of vocational behavior*, **10**: 316–340 (1977).

29 Nyman, K. & Raitasalo, R. *Absences from work and their determinants in Finland*. Helsinki, Research Institute for Social Security, 1978 (Report No. A 14/1978).

30 O'HANLON, J. F. Boredom: practical consequences and a theory. *Acta psychologica*, **49**: 53–82 (1981).

31 PEARLIN, L. & SCHOOLER, C. The structure of coping. *Journal of health and social behavior*, **19**: 2–21 (1978).

32 PERLMAN, B. & HARTMAN, E. A. Burnout: summary and future research. *Human relations*, **35**: 283–305 (1982).

33 PLANT, M. A. Occupations, drinking patterns and alcohol-related problems: conclusions from a follow-up study. *British journal of addiction to alcohol and other drugs*, **74**: 267–274 (1979).

34 PORTER, L. W. & STEERS, R. M. Organizational, work and personal factors in employee turnover and absenteeism, *Psychological bulletin*, **80**: 151–176 (1973).

35 ROMAN, P. M. & PRICE, H. M. The development of deviant drinking behavior: occupational risk factors. *Archives of environmental health*, **20**: 424–435 (1970).

36 SCHAR, M. ET AL. Stress and cardiovascular health: an international cooperative study—II The male population of a factory at Zürich. *Social science and medicine*, **7**: 585–603 (1973).

37 SEAMONDS, B. C. Stress factors and their effect on absenteeism in a corporate employee group. *Journal of occupational medicine*, **24**: 393–397 (1982).

38 SHARIT, J. & SALVENDY, G. Occupational stress: review and reappraisal. *Human factors*, **24**: 129–162 (1982).

39 SHIROM, A. ET AL. Job stresses and risk factors in coronary heart disease among five occupational categories in kibbutzim. *Social science and medicine*, **7**: 875–892 (1973).

40 SPERANDIO, J. C. Variation of operator's strategies and regulating effects on workload. *Ergonomics*, **14**: 571–577 (1971).

41 STRAYER, R. A study of employment and adjustment of 870 male alcoholics. *Quarterly journal of studies on alcohol*, **18**: 278–287 (1957).

42 UDRIS, I. Stress in Arbeitspsychologischer Sicht. [Stress from the point of view of industrial psychology.] In: Nitsch, J. R. *Stress: Theorien, Untersuchungen, Massnahmen.* Bern, Huber, 1981, pp. 391–440 (in German).

43 WELFORD, A. T. Stress and performance. *Ergonomics*, **16**: 567–580 (1973).

Chapter 5

# Neurophysiological reactions to stress

James F. O'Hanlon[1]

## Homeostasis and stress

Homeostasis, or the active maintenance of all the vital systems at the equilibrium levels conducive to optimum overall functioning, is common to all physiological concepts of stress. Each vital system may be considered to be a controlled system and every internal process that functions to maintain the homeostatic level of a particular system may be called a controlling mechanism. In the relatively simple case of human thermoregulation, the controlled system is body temperature, or more precisely brain temperature, and the controlling mechanisms include cardiac output, cutaneous blood flow, and sweating. A heat load, being an external stressor, will first activate the controlling mechanisms in the intact organism, and if their response is insufficient, eventually the controlled system itself. Throughout this process, the body can be said to be under stress—i.e., engaged in an effort to maintain thermal homeostasis however successfully. The physiological signs of increased cardiac output and increased cutaneous blood flow to provide greater heat loss by radiation, and sweating to provide heat loss by evaporation, are directly related to whole-body stress until those mechanisms are fully engaged or exhausted. The final rise in body temperature is a more significant sign of stress as it indicates a pathological reaction, but it is obviously less sensitive. Physiological or psychological reactions that are not specifically related to thermoregulation but are instead correlates of stress in general would be of little interest if they did not occasionally lead to another reaction, e.g., impairment of the motivation or ability to perform a critical task in the heat, or in the long term to secondary system failure through general neuroendocrine activation. Correlates of adaptive stress reactions are certainly the least reliable and the most difficult to interpret among the signs of stress produced by physical agents. Unfortunately, they are often the only measures available in the study of psychosocial stressors.

The psychosocial stressor is literally a creation of the human brain. That organ interprets perceived information in relation to information stored in the memory and an appraisal of its own capacity to overcome any threat the perceived information

[1] Department of Work and Organizational Psychology, University of Groningen, Groningen, Netherlands.

conveys. The external psychosocial hazard is an event, or combination of events, that the brain interprets as a threat to its ability to maintain a comfortable state of equilibrium and/or a desired mode of behaviour. These two factors may be viewed as constituting the controlled system. Satisfactory controlling mechanisms might be to act on the environment to eliminate the hazard, to reinterpret the hazard to reduce its threat, or in some way to increase the capacity of the brain to challenge the threat more effectively. Such reactions can be categorized under the single title "coping". Unlike physiological adaptation to physical stressors, coping depends on having the freedom, experience, and ability to select the most effective from among a number of possible means. Many situational and individual factors limit coping. Even when it is permitted within a given situation, the injudicious or merely unlucky selection of one particular means can intensify the threat and inhibit further attempts to cope.

The particular problems encountered in attempting to extend the classical homeostatic stress model to the concept of psychosocial stress can be summarized as follows:

(1) The "stressor" is internal and is mediated by the brain's interpretation of the information available about a given situation. Assuming the interpretation process is veridical, there must first exist a number of environmental factors, collectively defined as a hazard, which are fundamentally responsible for the internal reaction. Those factors are difficult to determine, owing to their multiplicity and to their varying ways of contributing to the reactions in different individuals. In short, the intensity of the stressor cannot be objectively defined, except possibly by its duration.

(2) The controlled systems are difficult to define and measure. Affective state and task performance are the most likely candidates, but if they are indeed the factors the brain seeks to control it is clear that in normal circumstances they cannot be simultaneously stabilized. In any practical situation there must be a compromise between efficiency and satisfaction. When there has to be a compromise both these controlled systems must depart from their normal state, with a continuation of disturbed homeostasis. It may prove necessary to develop a multidimensional scale to measure simultaneously the extent of departure of all controlled systems from their respective normal states, in order to estimate the amplitude of psychosocial stress. Establishing a scale would be difficult even if the systems to be controlled were known, and were responsive to psychosocial

hazards in the same way in all individuals. Because this is obviously not the case, the problem appears to be almost insoluble.

(3) Threat-controlling reactions are recognizable and their mere existence could be said to indicate the presence of psychosocial stress. However, as there are a great number and they can be ineffective or effective to widely differing degrees, none can be accepted as the single index of stress.

If it is not possible to define the stressor, the homeostatically controlled systems, or the controlling mechanisms, the value of the classical homeostatic model in the study of psychosocial stress is debatable. It would be logical to conclude that physical and psychosocial stress are fundamentally different phenomena were it not for one thing: their non-specific physiological correlates are often quite similar; therefore the long-term consequences of the non-specific reactions might be the same.

## Activation in the presence of a stressor

Cannon's famous "flight or fight" hypothesis pertains to man reacting to a threat just as it does to the presence of a physical agent causing pain (2). Simply stated, the hypothesis is that the stressor activates all the physiological systems that have the function of sustaining maximum strength, speed of response, and endurance in the skeletal musculature. This includes a rise in isometric muscle tension, a rise in cardiac output by means of increased myocardial strength and frequency of contraction, the contracture of arterioles in the cutaneous and mesenteric vascular beds, with redistribution of the blood flow in favour of the muscles, and the neurohumoral mobilization of hepatic glucose and free fatty acids and glycerol from the adipose tissue. Though not known to Cannon, this general activation perpetuates the brain's own state of arousal through a direct positive-feedback loop involving the adrenal medullary release of adrenaline which stimulates receptor sites in the posterior hypothalamus.

Clearly this is a beneficial reaction in a short-lasting emergency when physical exertion is necessary to overcome the threat. But frequently repeated episodes of this sort, or prolonged activation of the autonomic nervous system and various neuroendocrine systems, disturb homeostatis.

Selye recognized this in the late 1930s, primarily in relation to a particular component of general activation, namely the sustained high production of glucocorticoids (cortisol in man) from the adrenal cortex in response to adrenocorticotropic hormone from the adenohypophysis (anterior pituitary). The adaptive functions of cortisol include: inhibition of the spread of the inflammatory reaction that accompanies physical trauma; when the body is not receiving enough nourishment, transformation of protein metabolism from anabolism to catabolism, thereby ensuring blood glucose homeostasis through hepatic gluconeogenesis. In speculating on the other beneficial effects of elevated glucocorticoid production Selye saw few that were maladaptive. Subsequent research, however, has failed to confirm the beneficial effects, and has indeed revealed maladaptive effects.

## Maladaptive effects of long-term activation

It is popular, in current psychosomatic medicine and biological psychiatry, to theorize that all sorts of diseases may be ascribed to the chronic effects of over-activation in people exposed to psychosocial hazards, particularly those encountered at work. The pathophysiological mechanisms of psychosomatic disease are exceedingly obscure, however, and though several possible mechanisms have been suggested none has been observed experimentally.

A prolonged elevation of isometric muscle tension might be responsible for various diseases of the muscles, tendons, and joints, specifically the "occupational neck-shoulder-arm syndrome" (see page 43 below). Similarly, labile hypertension, due to episodic sympathetic nervous system activation, might affect the cardiovascular control system, causing essential hypertension and myocardial heart disease. Drastic fluctuations in the autonomic control of gastrointestinal blood flow, motility, and secretion might be responsible for peptic or duodenal ulceration, as well as chronic diarrhoea or constipation. The constant mobilization of carbohydrates or lipids, that are not subsequently metabolized by the skeletal musculature, might eventually give rise to other metabolic products—e.g., cholestorol—that accumulate in the intima of arteries to produce ateriosclerosis.

Many of the consequences to the systems of chronic over-activation, are mediated by an excessive production of sympathoadrenomedullary hormones, adrenaline and noradrenaline. These hormones may also directly affect the organs, e.g., the heart, causing arrhythmias, electrolyte imbalance, and even necrosis. Chronic activation of the adenohypophyseal-adrenocortical axis may produce local tissue damage, primarily as a result of the inhibition by cortisol of amino acid uptake by mucosal, skeletal muscle, skin, and lymphoid cells. A loss of resistance in the gastrointestinal mucosa to acid and proteolytic enzymes, muscle wasting, and diminished antibody production which increases the susceptibility to infection, are some of many possible results.

In short, numerous theories exist to explain practically every known pathophysiological reaction to prolonged activation as a consequence of psychosocial stress. Indeed, from such speculation it is difficult to imagine how anyone can avoid succumbing to disease under even normal everyday occupational stress. The fact is that psychosomatic disease is an exceptional result of psychosocial stress rather than the rule. Epidemiological research has failed so far to establish clearly the relationship between any psychosocial stress factor and any psychosomatic disease. This failure is almost certainly because compensatory psychological and physiological processes operate during periods of intense activation and during interpolated recovery periods. In a relatively few individuals, those processes are insufficient and they do eventually succumb, though in a variety of ways. To establish the causal link between psychosocial hazards and psychosomatic disease it is necessary to define the more prevalent physiological reactions that can be monitored in individuals exposed to supposed hazards. Not every physiological reaction is important in this respect, only those that can logically be considered precursors to disease.

## Physiological monitoring to determine the occurrence of occupational psychosocial stress

Neuroendocrinological indices of psychosocial stress have been the most widely studied in working populations and most of them are discussed at length in Chapter 6. In this section, therefore, discussion will concentrate on continuous or periodic measures of physiological reactions that reflect cardiovascular, locomotor, gastrointestinal, or central nervous system functions. The scope of the review will be further limited to a few representative studies of "normal" workers and will not include individuals diagnosed as suffering from a medical disorder. Finally, only the results of studies on workers exposed to suspected psychosocial hazards will be included, not those on workers exposed to physical or chemical hazards that are definitely known to cause disease.

## Cardiovascular functions

The most extensively studied population in relation to the cardiovascular effects of psychosocial stress is that comprising the air traffic controllers in the USA. In the early 1970s it became evident that grave responsibility, periodically heavy mental work-load, irregular work/rest cycles, and other factors, create a potentially dangerous situation for the controllers' health and for the safety of airline crews and passengers. A 5-year empirical study was sponsored by the United States Federal Aviation Administration (FAA) which culminated in a report published in 1978 (8). The study included an epidemiological survey of morbidity among the controllers, and the monitoring of cardiovascular parameters as they worked. The former was successful in demonstrating that the incidence of essential hypertension was 3–4 times greater in the controllers than in comparable workers in other occupations. Repeated measure-ment of the blood pressure of 382 controllers during all phases of their work revealed that on the whole they did not have regular labile hypertension: a change from low to high work-load was accompanied by modest, if statistically significant, elevation in both systolic and diastolic pressures. However, 36 controllers did become hypertensive according to the FAA's criteria (systolic pressure $\geqslant 140$ mmHg (18.7 kPa); diastolic pressure $\geqslant 90$ mmHg (12 kPa) during the course of the study. Compared with the controllers who remained normotensive, the hypertensive group showed not only higher average systolic and diastolic pressures during work, but also much greater changes in systolic pressure due to variations in work-load. The work-load, a psychosocial stress factor, was clearly a contributor to the development of hypertension in the particularly susceptible individuals.

Insufficient justice is done to this extensive investigation by the brief description given above. To be sure it was not perfect. It might have been important, for example, to determine what the controllers' blood pressure was on their free days, or when they returned from vacation. Nevertheless, it probably provides the best record available of a chain of events that begins with an occupational psychosocial hazard and ends in disease. It also demonstrates the difficulties to be encountered in trying to establish the relationship; the sample must be very large indeed for it to include enough susceptible individuals. The period during which the workers are observed must be long enough to allow disease to develop in the susceptible individuals (3 years in the investigation described). The unflagging support of the research sponsor, the workers, their union representatives, and their management is a mandatory requirement. A diligent analysis of factors that might produce stress and the judicious application of sophisticated methodology in order to assess their relevant physiological effects are also essential from the beginning. It is, therefore, little wonder that this type of approach has seldom been applied. Other attempts have been less comprehensive and have, as a consequence, provided less convincing results.

## Locomotor functions

A particular effect of work stress called the "occupational neck-shoulder-arm syndrome" or "cervicobrachial disorder" has been described by Japanese physicians. It comprises muscular fatigue; stiffness and pain in the neck, back, and forearms; coldness and hypo-aesthesia (or paraesthesia) of the hands; headache; insomnia; and secondary emotional complaints (6). These symptoms occur particularly in office workers who operate machines—e.g., typewriters—and in industrial assemblers who are also required to maintain the same posture while engaged in repetitive motor activity. The restriction of gross movement, the prolonged isometric tension required to support the upper limbs, and an excessive and monotonous mental load, resulting in general activation, are seen as the primary causal factors. Slow recovery from overactivation, which affects sleep and interpersonal relationships outside work, causes further complications. Severe neurological disorders—e.g., neuritis or radiculitis—and various diseases of the muscles, tendons, and joints are among the suggested ultimate consequences.

If this syndrome is a combination of physical and psychosocial stress involving the neuromuscular system, a first sign should be increased isometric tension in the muscle groups maintaining posture but not those involved in the manipulations demanded by the task. Moreover, though the physical load is light and remains constant, tension might be expected to rise with increasing mental work-load, as the day progresses, or as the task is repeated as a result of fatigue and increasing stress. This could be measured objectively by means of continuous electromyographic (EMG) recording.

This approach was used in two studies involving industrial assemblers (4, 5). It was found that neck muscle tension varied in direct correlation with mental work-load and generally increased as the day progressed while the workers performed the same repetitive tasks. In addition to the direct effects of tension, the workers suffered from a variety of affections associated with chronic overactivation; so the results suggest that the increased tension through mental work-load may contribute to the various affections. It seems that both the local and the general components of such a state can be inferred from monitoring. An extension of this approach to other working populations and the association of EMG monitoring and recognized medical signs of neuromuscular disease would be required to establish the general applicability and importance of these findings.

## Central nervous system functions

There is no accepted definition of what constitutes homeostasis in the central nervous system. Possibly the myriad functions of the central nervous system preclude a single definition. If this is so, it will never be possible to determine from physiological measurements when the central nervous system is in a state of disturbed homeostasis or stress, at least in its normal waking condition. However, it is now commonly accepted that the activation level of the cerebral cortex determines the efficiency of its information processing function; that there exists an optimum intermediate level of activation for any individual or task; and even that the brain itself is sensitive to its own activation, or can become so, given appropriate sensory stimulation from the results of the behaviour it controls. These premises have led many theorists to suppose that the concept of homeostasis applies to the brain's control of its own activation level by both intrinsic and extrinsic mechanisms; and that mental effort is associated with the maintenance of activation homeostasis in situations where it is not sustained naturally. When the task to be performed is really critical an anticipated failure of effort can be a threat capable of evoking anxiety and peripheral activation. For example, it was observed that anaesthesiologists assisting at long surgical operations frequently succumb to monotony, with lapses in vigilance (1). As they become aware of such episodes they experience "mini-panics" until they have ascertained that nothing has gone amiss during the lapse. The mini-panics were observable in heart-rate recordings taken from the anaesthesiologists during their work and were confirmed by them afterwards. It can reasonably be assumed that they attempted to forestall their lapses by effort and experienced stress after each one and in anticipation of the next.

Other occupational groups whose tasks are similarly monotonous, and whose performance failures would also be crucial, experience an extraordinarily high incidence of psychosomatic disease (7). The hazard to health is not simply that of central nervous system deactivation in a monotonous working environment, it is that deactivation poses a significant threat to meeting an accepted responsibility in carrying out a task.

Current techniques permit the continuous electroencephalographic (EEG) recording of electrocortical activity in workers in their normal occupations. One of the first attempts at EEG monitoring in industrial workers—electronic assemblers—over an entire day showed a surprisingly high number of deactivation episodes, marked by the occurrence of theta rhythm (3). Sometimes such episodes have been called "micro-sleeps". Were the physiological monitoring of "micro-sleeps" and "mini-panics", together with other signs of whole-body stress in workers performing critical, monotonous tasks, to be studied in relation to one another, the genesis of a very prevalent form of psychosocial stress might well be discovered.

## Gastrointestinal functions

Although a beginning has been made in the large-scale physiological monitoring of cardiovascular, locomotor, and central nervous system disorders in workers in certain high-stress occupations, the investigation of functional disorders in other systems in occupational settings has so far been beyond the capacity of research workers, for technical or economic reasons, or both. In no system is this failure so apparent as the gastrointestinal, particularly in regard to such disorders as gastritis and peptic or duodenal ulcer, which are recognized as classic stress diseases. Many of the contributing factors hypothesized, such as a reduced stomach blood flow during periods of sympathetic nervous system activation, are seemingly impossible to measure in living persons. Other factors, such as stomach acidity and motility, can be monitored, but only with great difficulty and discomfort to the subject. Nevertheless, at least one likely correlate of gastric malfunction, the plasma pepsinogen concentration, can be readily measured. Apparently no investigator has as yet attempted to monitor plasma pepsinogen concentrations in workers from an occupational group known to suffer from frequent gastrointestinal disorders. This should be done, while continuing the search for better methods to monitor gastrointestinal functions.

## Summary and conclusion

Homeostasis remains the key to the stress concept, which now extends into the area of psychosocial stress generated by the human brain as it analyses a situation in relation to its coping resources and appreciates a threat. The classic concept in physiology of controlled systems and controlling mechanisms for adaptation, and of non-specific activation during incomplete adaptation, still possess considerable heuristic value. Though some theories have been postulated (9), investigators are at a loss to identify the homeostatically controlled parameter whose departure from equilibrium determines the amplitude of psychosocial stress; it is not possible to go much further towards a definition of the processes for controlling psychosocial stress than to list a large number of psychological, behavioural, and social reactions under the label "coping". Yet the signs of persistent overactivation that accompany complaints of stress cannot be readily explained without reference to the classic homeostatic model. Moreover, the same signs are indicative of processes within the body that can become pathogenic over a long period. Physiological monitoring to detect those signs at an early stage might be useful in order to confirm what man experiences in relation to stress and to indicate when he is insensitive to stress. More importantly perhaps, the simultaneous monitoring of enough of the right signs might indicate what disorder is likely to occur first, allowing the timely and specific application of countermeasures.

The effectiveness of this approach in an occupational environment has not been demonstrated often enough. All the serious studies have been lengthy and costly and whether the effect of their findings on local, or industry-wide, working conditions has justified the effort is questionable. However, the effort has barely started. The major success has been in demonstrating the feasibility of physiological monitoring; an approach that, in the study of air traffic controllers, has already proved to have predictive validity. It may be expected that not only will these pioneering activities be replicated and extended but, as better methods become available from basic research, new ways will be found of physiologically monitoring the reactions of workers in their occupational settings.

## References

1 BRANTON, P. & OSBORNE, D. H. A behavioural study of anaesthetists at work. In: Oborne, D. J. et al., ed. *Research in psychology and medicine.* London, Academic Press, 1979, Vol. 1, pp. 434–441.

2 CANNON, W. B. *Bodily changes in pain, hunger, fear and rage,* 2nd edition. New York, Apploton, 1020.

3 CHELIOUT, F. ET AL. Rythme thèta postérieur au cours de la veille active chez l'homme. *Revue d'electroencephalographie et de neurophysiologie clinique,* **9**: 52–57 (1979).

4 LAVILLE, A. Cadence de travail et posture. *Le travail humain,* **31** (1–2): 73–94 (1968).

5 LAVILLE, A. ET AL. *Conséquences du travail répétitive sous cadence sur la santé des travailleurs et des accidents.* Paris, Laboratoire de Physiologie du Travail et d'Ergonomie, 1973 (Rapport No. 29 bis).

6 MAEDA, K. Occupational cervicobrachial disorder in assembly plant. *The Kurume medical journal,* **22**: 231–239 (1975).

7 O'HANLON, J. F. Boredom: practical consequences and a theory. *Acta psychologica,* **49**: 53–82 (1981).

8 ROSE, R. M. ET AL. *Air traffic controller health change study.* Springfield, VA, United States National Technical Information Service, 1978 (Office of Aviation Medicine Report No. FAA-AM-78-39).

9 STAGNER, R. Homeostasis, discrepancy, dissonance: a theory of motives and motivation. *Motivation and emotion,* **1**: 103–138 (1977).

Chapter 6

# Metabolic and neurohormonal reactions to occupational stress

Meglena Daleva[1]

## Introduction

The state of the body is determined by a great variety of functional processes, biophysical, biochemical, neurophysiological, and neurohormonal. These processes aim at maintaining the body's homeostatic activity, regardless of whatever environmental changes occur. Adaptation to the external environment is a principal area of study for psychologists, biologists, physiologists, and sociologists and in this modern age their investigations are enhanced by the results of rapid scientific and technical progress.

The number of people engaged in different industrial spheres, in transport, and in education is increasing; new professions have been recognized; man has reached the ocean depths and outer space. Contemporary life thus presents new challenges that often do not correspond to man's capabilities for adaptation.

The ability of the body to exist, and preserve its internal stability under different environmental conditions is, to a large extent, based on the stability and dynamism of its physiological functions.

## The neurohormonal mechanism

It is currently accepted that complex mechanisms are activated in stress situations. In all such situations it is the hypothalamus that activates the body's protective mechanisms. There are two ways in which the hypothalamus/pituitary axis is first stimulated to react to stress and then acts to bring about hormone regulation. The first is neural, the hypothalamus being stimulated from higher centres to regulate the reaction to stress whatever its intensity and character; the second is humoral, the pituitary gland constantly interacting with the adrenal, thyroid, and sexual glands. The hypothalamus reacts quickly to every signal by secreting the necessary substances in the blood and stimulating the corresponding endocrine gland (Fig. 2). The highest coordinating and regulating centres of the autonomic nervous and endocrine systems, which react to even the smallest disturbances, are situated in the hypothalamus, which, in turn,

[1] Institute of Hygiene and Occupational Health, Sofia, Bulgaria.

Fig. 2. A simplified diagram of the main hypothalamic and pituitary control mechanisms

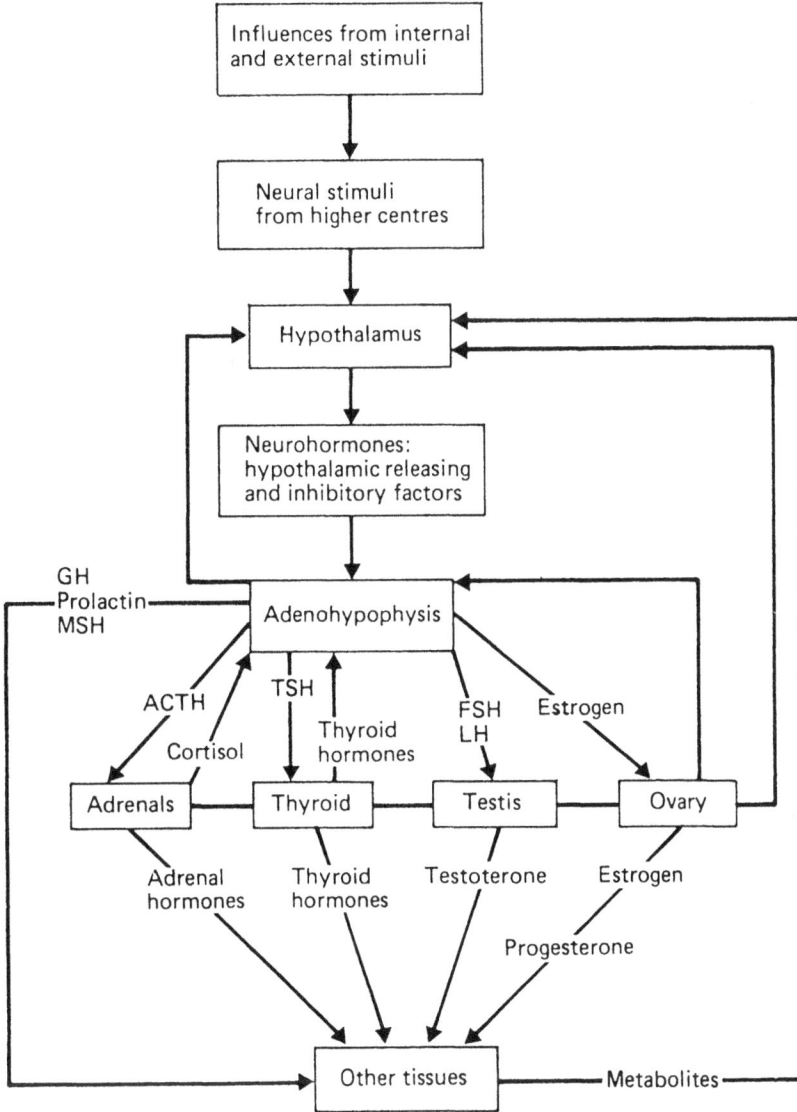

SYMBOLS:

GH    = growth hormone.
MSH   = melanocyte stimulating hormone.
ACTH  = adrenocorticotropic hormone.
LH    = luteinizing hormone.
TSH   = thyrotrophin releasing hormone.
FSH   = follicle stimulating hormone.

WHO 851813

receives messages from the brain cortex. Several parts of the endocrine system that are controlled by the hypothalamus react simultaneously in stressful situations. Among them, two are of special importance: the sympathetic-adrenal-medullary system secreting catecholamines; and the pituitary-adrenal-cortical system secreting corticosteroids.

The immediate neuroendocrine response to stress leads to the activation of the sympathetic-adrenal-medullary system. The most important adrenergic substances secreted are the catecholamines—adrenaline and noradrenaline. Adrenaline is the main hormone of the adrenal medulla; noradrenaline is its immediate precursor. Noradrenaline also functions as an adrenal medullary hormone and as a transmitter in the central and sympathetic nervous systems.

The catecholamines are among the most important regulators in the adaptation mechanisms of the body. They enable it to convert quickly and adequately from a state of rest to a state of arousal and to maintain that state for a long time. They elicit a wide variety of effects as is illustrated by the differences in response produced by adrenaline and noradrenaline presented in Table 1.

The importance of the catecholamines as body regulators is based on their ability to increase and intensify its metabolism. They stimulate the decomposition of glycogen and lipids, cause the accumulation of glucose in the blood, activate the oxidation of fatty acids, stimulate heart and muscle activity, stimulate the central nervous system, and influence the protective and

Table 1. Effects of the intravenous infusion of adrenaline and noradrenaline in man

| Index | Adrenaline | Noradrenaline |
|---|---|---|
| Heart rate | + | − |
| Cardiac output | + + + | 0, − |
| Systolic blood pressure | + + + | + + + |
| Diastolic blood pressure | +, 0 | + + |
| Total peripheral resistance | − | + + + |
| Oxygen consumption | + + | 0, + |
| Blood glucose | + + + | 0, + |
| Blood lactate | + + + | 0, + |
| Blood nonesterified fatty acids | + + + | + + + |
| Central nervous system action | + | 0 |
| Eosinopenic response | + | 0 |

+ = increase; 0 = no change; + = decrease.
*Source:* Reference *31*.

immunological processes and the other adaptation mechanism, i.e., the hypothalamic-hypophysis-adrenal-cortical system.

In a discussion of the biochemical mechanisms of reaction to stress in occupations and circumstances connected with emotional strain, e.g., in drivers, public performers, parachutists, and alpinists, the part played by catecholamines is emphasized as well as their role in the genesis of cardiovascular disease (13). Under the influence of an increased production of adrenaline and noradrenaline in stress situations, triglycerides are subjected to lipolysis, which causes the level of free fatty acids in the blood to increase; free fatty acids deposited on the arterial walls in conditions of increased glycaemia are converted into triglycerides. This underlies the process of atherogenesis. Atherogenesis is conditioned by numerous genetic and environmental factors, such as diabetes, cigarette-smoking, lack of exercise, obesity, or a high content of saturated fatty acids in the diet.

The three following types of steroid hormone that act in stress situations are released from the adrenal cortex:

(a) *Mineralocorticoids* which control the mineral concentration in the body, especially the contents of the sodium ions. They stimulate kidney activity during sodium reabsorption from the urine and its retention in the tissues. Although the release of mineral corticosteroids is activated insignificantly by adrenocorticotropic hormone it is controlled by the renin–angiotension mechanism, which is dependent on the sodium level in the blood and on arterial pressure.

(b) *Glucocorticoids*, the most important of which is cortisol. Among the basic functions of the glucocorticoids at normal concentration is the facilitation of water excretion through the kidneys, and the maintenace of normal arterial pressure. They also supply the body with energy in order to increase its resistance. At lower concentrations they affect the infectious processes and protein synthesis, increase the rate of calcium and phosphate release from the kidneys, and increase the blood sugar level. The glucocorticoids are released under the influence of adrenocorticotropic hormone.

(c) *Androgens* which are important hormones in both sexes though they are regarded as the male sexual hormones.

More than 50 years have passed since the results of studies on the activation of the adrenal medulla and the adrenal cortex in stress situations were first published. Since that time there have been a number of studies on the various biochemical indicators of stress, carried out in the laboratory, in animals, and in man, the results of which have also been published. The many important questions that arose during the studies led to the creation of accurate and highly sensitive methods for determining the biochemical indicators of stress in tissues and body liquids—spectrophotometric, fluorometric, chromatographic, radioisotopic, and radioenzymatic methods. Through fluorescent and immunological techniques it became possible to determine the noradrenergic, dopaminergic, and adrenergic functions of the brain. The development of microdissection and histochemical techniques and extremely sensitive methods of measuring catecholamine and other biogenic amines in the small brain areas led to the discovery that stress has a selective influence on specific brain nuclei.

The considerable progress made in the field of neurobiology in the past decade is due to the discovery of many other transmitters—amino acids and neuropeptides—that also play a role in reaction to stress.

## The circadian rhythms

At this point it becomes necessary to consider some of the problems connected with the circadian rhythms, since they are important in evaluating the hormonal changes caused by the working activity of man. As a current procedure in work physiology, urine and blood are often collected at short intervals and at different hours of the day. This raises the question of what are normal values and how do they vary; an understanding of the normal diurnal variations in hormonal excretion is required.

The term biological rhythms in the broader sense involves periodic and cyclic variations differing from one another in character and duration. They include the circadian rhythms, and biological phenomena occurring monthly, seasonally, annually, and perennially. The circadian rhythms are the best known biological rhythms; their periodicity is about 24 hours.

A study of the variables that are subject to a circadian rhythm and the degree of influence of external factors over these rhythms provides an opportunity to evaluate degrees of stress in different situations, both simulated in the laboratory and in real life, where the character and intensity of work are contributory causes.

The "synchronizers" in the environment may influence the pattern of hormonal excretion (10). These synchronizers include the regular changes from day to night, temperature variations, and the periodicity of meals. Other types of synchronizer are of a social nature and include those relating to working conditions, e.g., shift work with alternating periods of night work and day work, alternating periods of rest and work, long working hours with a high degree of responsibility, physical load, and monotonous tasks (3, 18, 33, 34, 45, 49, 63, 64). Work involving physical or psychological strain, or shift work, modify the diurnal variations of the physiological functions. Various synchronizers influence the parameters of the rhythm. On the basis of changes caused by work in the diurnal rhythms of the neurohormonal functions, inferences can be made about work-load and the level of adaptation to it.

In the earliest studies of adrenaline and noradrenaline secretion clear differences were shown between sleeping and waking, with higher excretion values during the day when the subjects were awake (1). In a number of recent studies urine samples have been collected systematically, enabling statistical analysis of the circadian rhythms. Clear circadian rhythms were found in relation to the secretion rates of catecholamines, with the highest excretion values in the early afternoon and the lowest during the night (2, 19).

Studies on corticosteroid (cortisol) secretion in which 4–6 hours elapsed between the collection of samples yielded information on circadian variations in the amount of corticosteroids in the plasma, with the excretion values in the morning before waking and a progressive decrease in hormone level during the day. By means of urine and blood samples taken at short intervals it has been possible to study circadian changes in the plasma corticosteroids in detail. It was found that the quantity of adrenocorticotropic hormone and plasma cortisol in a healthy person increased early in the morning (approximately 40% of the 24-hour total) and decreased as the day progressed (only 5% secreted during the evening) (4).

In another study it was found that the amount of cortisol excreted in the urine reflects variations in the secretion of plasma cortisol. The highest excretion values were detected between 08 h 00 and 10 h 00 (12).

The excretion level of corticosteroid metabolites is lower during the night than during the day. The data on the dynamics of excretion during the day are somewhat inconclusive. In some studies high values were detected in the morning and lower ones in the afternoon. In others no considerable difference was found between the morning and the late afternoon (61); and in still

others the corticosteroid metabolites as well as the catecholamine metabolites showed daily variations (*63*).

## Individual variations in the excretion of neurohormones

In spite of a certain homogeneity in the excretion rates of
neurohormones among population samples, individuals may
differ considerably in their response patterns.

In a study published in 1980 persons classified as Type A and
Type B (see Chapter 13) were compared in terms of psycho-
physiological arousal, during periods of rest and following
strenuous mental activity. Judging by catecholamine and
cortisol excretion levels and heart rate, the Type A persons were
either equally aroused or more aroused during rest than after
strenuous mental activity, whereas the Type B persons were
more aroused during strenuous mental activity than during rest.
It is postulated that a continuously elevated hormone excretion
level adds to the health risks associated with Type A behaviour
(*27*).

It has been shown that people who prefer to work in the
evening excrete more adrenaline in the urine in the evening
than in the morning, compared with people who prefer to work
in the morning, who excrete more adrenaline in the morning
even when they are at rest. It was found that catecholamine
values are considerably higher in people working faster, more
accurately, and "easily" (*23*).

Many investigations have been carried out on sex differences
in relation to stress. It was found that there were no differences
between men and women in cortisol and catecholamine excretion
rates during rest periods. Under stress, the reaction
of the sympathetic-adrenal-medullary and pituitary-adrenal-
cortical systems is considerably stronger in men than in women.
Such sexual differences were detected in adrenaline, cortisol,
and 4-hydroxy-3-methoxyphenylethyleneglycol (MHPG—a metabolite that expresses noradrenaline synthesis and turnover in
the brain) excretion rates (*15, 26*).

## Hormonal changes during work

The changes that take place in the sympathetic-adrenal-
medullary and pituitary-adrenal-cortical systems are indicators
that characterize the reaction of the body to the influences of
work. When analysing the changes in these biochemical systems

it must be kept in mind that although man is more or less adapted to work there are various factors influencing him continuously and the degree of stress he experiences is a combination of factors related to content, organization, and environmental conditions. Some of the basic factors that influence the activity of the two systems are discussed below.

## The influence of the physical work-load

Much evidence leads to the conclusion that the reactions of the sympathetic-adrenal-medullary and pituitary-adrenal-cortical systems vary according to the nature and intensity of physical load. Being in a state of stress may increase the effect of the physical work-load. It has been shown that physical over-activity, such as exercise and an increase in loading on the bicycle ergometer, considerably enhances cortisol, adrenaline, and noradrenaline secretion levels as well as the activity of dopamine $\beta$-hydroxylase enzyme (EC 1.14.17.1). Studies also show that corticosteriod and catecholamine secretion levels increase in people doing hard physical work (11, 32, 43, 48, 51, 54, 65).

## The influence of psychological stress

The influence of psychological stress on the adaptive systems has been investigated extensively and a great number of psychoendocrinological studies have been published in the last 20 years. This is due not only to the progress made in the fields of biochemistry and electrophysiology but also to improved methods of analysing changes that take place in human behaviour.

Among the numerous studies on changes in corticosteroid and catecholamine secretion levels as a response to stress at work are some that relate to the following: pressure due to shortage of time for performing a specific task, including responsibility as it pertains to such people as pilots and air traffic controllers (21, 41), locomotive engineers (47, 60), and firemen (20, 40); technically advanced work processes (24); computerized work (39); workers in jobs that require adaptation (58); piece-work and work on the assembly line (59); opera singers and members of orchestras during training and performance (42); increased responsibility at work (19); teachers with an intensive teaching programme (16); monotonous work (58). In a study published in 1980 it was found that what people do and what they experience are reflected in urinary catecholamine excretion rates, particularly adrenaline excretion, which are higher when they are at work than when they are involved in leisure pursuits (37). Adrenaline secretion is thus elevated in conditions of continuous mental stress; it is extremely important in mobilizing the body's protective mechanisms in unexpected situations (22). Noradrenaline secretion is also modified by emotions, though to a lesser extent than adrenaline. Feelings of boredom and physical tiredness are associated with low adrenaline excretion rates, while mental tiredness is related to high adrenaline excretion rates (53). Noradrenaline is influenced to a greater degree by the physical work-load. Noradrenaline is also influenced by the posture of the worker and by physical factors in the working environment.

Activation of the secretion of catecholamines as a reaction to stress facilitates adaptation to the stressors in the environment (22). A positive correlation was found between rate of adrenaline excretion, social adaptation, and emotional stability (38). The secretion of catecholamines is important in ensuring flexibility in adapting to different situations. There are enough data to show, however, that continuous or redundant stimulation of the sympathetic-adrenal-medullary system may cause unfavourable bodily changes (14).

## Exposure to physical factors in the working environment

The functions of the pituitary-adrenal-cortical and sympathetic-adrenal-medullary systems during work are affected by the degree of physical loading, by physical factors combined with factors in the social working environment, and by work content and organization.

Exposure to noise and vibration has an influence on the corticosteroid excretion rates in urine in both man and animals (66). Exposure to excessive noise in man leads to increased corticosteroid excretion rates in both the blood and the urine. Both increases and decreases in rates of 17-ketosteroid excretion have been observed. Continuous exposure to noise leads to a decrease in the 17-ketosteroid excretion rate. Subjects exposed to high levels of noise at various frequencies show elevated corticosteroid excretion rates (6).

The effect of noise upon catecholamine secretion usually leads to an increased excretion rate in the urine. In a study published in 1977 the combined effect of task performance level and noise level resulted in a higher excretion of adrenaline than noradrenaline, while the corticosteroid excretion rate hardly altered (25). A significantly high content of vanillylmandelic acid (VMA—a catecholamine metabolite) was found in the urine of workers at a noisy (98–127dB; 40–1000Hz) construction workshop (5). In an investigation in the laboratory of the effect of acute noise (80dB) on corticosteroid, adrenaline, and noradrenaline excretion rates, significantly increased adrenaline excretion rates were found; noradrenaline excretion rates were also increased but to a lesser degree; no change was found in corticosteroid excretion rates (56). The increased adrenaline excretion rates in interpreters exposed to additional sound stimuli of 60dB, 70dB, and 85dB were found to be proportional to the noise intensity and the noradrenaline excretion rates increased significantly at 85dB (62). In a study of workers continuously exposed to industrial-type noise and total body vibration clearcut increases in noradrenaline excretion rates were found, which seem to have been related inversely to other functional disorders (46). High rates of catecholamine and corticosteroid excretion were found in tractor drivers, as a result of the combined effect of various factors, such as noise, total body vibration, and physical activity (18).

## Shift work

Constant efforts to introduce measures for uninterrupted work in different industrial fields has led to a wider utilization of shift work all over the world. Shift work creates social and health problems. Of special concern is the extent of the body's capability to adapt itself to night-work conditions; man functions less actively during the night and a considerably higher intensity in adaptive mechanisms is required to ensure a daytime level of activity. The pituitary-adrenal-cortical and sympathetic-adrenal-medullary systems are both important in the process of adapting to shift work and working at night. Many investigations have been carried out on the changes that take place in hormonal excretion rates during shift work and night work. Perhaps the most important physiological problem is related to the recovery of the physiological functions after a change in working and sleeping cycles. The basic facts concerning this problem are well known, having been ascertained in studies in animals (8), and in man in controlled environmental conditions—e.g., in isolated places or in caves—

where the subject is totally removed from all contact with the outside world (9, 30, 52).

The main conclusion that can be drawn from such studies is that man, in the same way as animals, has innate circadian rhythms that are influenced by *zeitgebers* (7) or synchronizers (35). It has been shown many times that, for animals, light is the most important synchronizer. In man, however, the dominant synchronizers are the social environment and an awareness of clock time. Light is also a synchronizer in man but a relatively weak one (8).

Shift work produces a situation where there is an alteration in waking and sleeping phases without a corresponding alteration in the phasing of the dominant social synchronizers. This is where the main problem lies. The same types of functional change occur in the first few days after a transmeridional air journey. Thus, the results of studies in connection with transmeridional flights provide good models for an understanding of the problems of shift work. Other models that enhance understanding are provided by the results of experimental longevity studies in insects involving phase-shifted light/dark schedules (36). These studies suggest that different light/dark cycles have different effects on longevity. Although they are related to some lower species and cannot be extrapolated directly to shift work, they may provide guidance in designing studies in man.

There is a need for laboratory and field studies on specific problems related to the recovery of physiological functions disturbed by shift work. Such status would have to be specially designed since post hoc studies are not sufficient. Several specially designed studies have been carried out in recent years (17, 50, 55). Different types of shift system were examined by measuring the various physiological and psychological functions of the workers involved over periods of time ranging from 1 day to several weeks. In the laboratory experiments, attempts were made to simulate real-life situations as far as possible—i.e., "open-door" experiments without social isolation and with real or simulated work. It was observed that in systems with night shifts the circadian rhythm of the body temperature was not significantly altered. However, in experiments which involved continual night work for periods of 1–3 weeks alterations in minimum temperature were observed. The recovery rate of body temperature was measured in persons who had taken transmeridional flights (29); it is not known at present whether other bodily functions recover at similar rates.

The biochemical effects of shift work have been studied mainly from the point of view of occupational stress (28, 44). What is needed is much more research on catecholamines and on serotonin in connection with shift work. From a medical point of view it is the physiological problems of shift work that are seen to be the most important. The shift workers themselves, however, regard the psychosocial problems as being of more consequence, to the extent that their family and social life are disrupted. Thus, although it may be possible to plan a shift system in which the possibility of resulting physiological problems is minimized, it may not be acceptable to the worker. Research could lead to the preparation of guiding principles which would aid managers in industry to construct shift schedules that would take into account basic physiological requirements yet, at the same time, respect the workers' social needs.

## References

1 ÅKERSTEDT, T. Åltered sleep/wake patterns and circadian rhythms. *Acta physiologica Scandinavica*, Suppl. 469, 13–16 (1979).

2 ÅKERSTEDT, T. & LEVI, L. Circadian rhythms in secretion of cortisol, adrenaline and noradrenaline. *European journal of clinical investigations*, **8**: 57–58 (1978).

3 ÅKERSTEDT, T. ET AL. Field studies of shiftwork: II. Temporal patterns in psycho-physiological activation in workers alternating between night and day work. *Ergonomics*, **20**: 621–631 (1977).

4 ANDERS, T. F. Biological rhythms in development. *Psychosomatic medicine*, **44**: 61–72 (1982).

5 ANITESCO, C. & CONTULESCO, A. Etude de l'influence de bruit et des vibrations sur le comportement des catécholamines dans l'agression sonore vibratoire industrielle. *Archives des maladies professionnelles, de médicine du travail et de sécurité sociale*, **33**: 365–372 (1972).

6 ARGUELLES, A. E. Pituitary adrenal stimulation by sound of different frequencies. *Journal of clinical endocrinology and metabolism*, **22**: 846–852 (1962).

7 ASCHOFF, J. Zeitgeber der tierischen Tagesperiodik. [Zeitgebers of animal circadian periodicity.] *Naturwissenschaften*, **41**: 49 (1954) (in German).

8 ASCHOFF, J. ET AL. Re-entrainment of circadian rhythms after phase-shifts of the zeitgeber. *Chronobiologia*, **II**: 23–78 (1975).

9 ASCHOFF, J. & WEVER, R. Spontanperiodik des Menschen bei Ausschluss aller Zeitgeber. [Spontaneous periodicity in man when all zeitgebers are excluded.] *Naturwissenshaften*, **49**: 337–345 (1962) (in German)

10 ASCHOFF, J. & WEVER, R. Uber Reproduzierbarkeit circadianer Rhythmen beim

Menschen. [On the reproduction of circadian rhythm in man.] *Klinische Wochenschrift*, **58**: 323–335 (1980) (in German with an abstract in English).

11 BALAZOVJECH, T. ET AL. Effect of work load on plasma dopamine-$\beta$-hydroxylase activity and cortisol in patients with essential hypertension. In: Usdin, E. et al., ed. *Catecholamines and stress*. Oxford, Pergamon Press, 1976, pp. 549–556.

12 BRANDENBERGER, G. & FOLLENIUS, M. Variations diurnes de la cortisolémie de la glycémie et du cortisol libre urinaire chez l'homme au repos. *Journal de physiologie*, **66**: 271–282 (1973).

13 CARRUTHERS, M. Biochemical responses to environmental stress. In: Michelson, W. H., ed. *Man and his urban environment: a sociological approach*. Oxford, Addison-Wesley, 1976, pp. 244–273.

14 COGAN, B. M. [Stress and adaptation.] In: *Biologii*. Moscow, Znanie, 1980, pp. 64–68 (in Russian).

15 COLLINS, A. & FRANKENHAEUSER, M. Stress responses in male and female engineering students. *Journal of human stress*, **4**: 43–48 (1978).

16 DALEVA, M. & HADŽIOLOVA, I. [Biochemical aspects of the strain of teachers' work.] *Problemi na higienata*, **V**: 28–36 (1980) (in Bulgarian with a summary in English).

17 DALEVA, M. ET AL. Physiological changes in operators during shift work. In: Swensson, A., ed. *Night and shift work*. Stockholm, Studia laboris et salutis, Report No. 11, 1972, pp. 26–32.

18 DALEVA, M. ET AL. Changes in the excretion of corticosteroids and catecholamines in tractor drivers. *International archives of occupational and environmental health*, **49**: 345–352 (1982).

19 DALEVA, M. ET AL. Changes in circadian rhythm of catecholamine excretion in shift workers under neuroemotional stress. In: Usdin, E. et al., ed. *Catecholamines and stress: recent advances*. Amsterdam, London, and New York, Elsevier, 1980, pp. 471–476 (Developments in Neuroscience Series, No. 8).

20 DUTTON, L. M. ET AL. Stress levels of ambulance paramedics and fire fighters. *Journal of occupational medicine*, **20**: 111–115 (1978).

21 ELIOT, R. S. ET AL. Influence of environmental stress on pathogenesis of sudden cardiac death. *Federation proceedings*, **36**: 1719–1724 (1977).

22 FRANKENHAEUSER, M. Experimental approaches to the study of catecholamines and emotion. In: Levi, L. ed. *Emotions: their parameters and measurement*. New York, Raven Press, 1975, pp. 209–234.

23 FRANKENHAEUSER, M. & ANDERSSON, K. Note on interaction between cognitive and endocrine functions. *Perceptual and motor skills*, **38**: 557–558 (1974).

24 FRANKENHAEUSER, M. & GARDELL, B. Underload and overload in working life: a multidisciplinary approach. *Journal of human stress*, **2**: 35–46 (1976).

25 FRANKENHAEUSER, M. & LUNDBERG, U. The influence of cognitive set on performance and arousal under different noise loads. *Motivation and emotion*, **1**: 139–149 (1977).

26 FRANKENHAEUSER, M. & LUNDBERG, U. Psychoneuroendocrine aspects of effort and distress as modified by personal control. In: Bachman, W. et al., ed. *Mental load and*

*stress in activity*. Berlin, GDR, Verlag der Wissenschaften, 1980, p. 43. (Abstracts in Proceedings of the XXIInd International Congress of Psychology, Leipzig, 6–12 July 1980).

27  FRANKENHAEUSER, M. ET AL. Note on arousing type-A persons by depriving them of work. *Journal of psychosomatic research*, **24**: 45–47 (1980).

28  FRÖBERG, J. E. & ÅKERSTEDT, T. *Night and shift work effects on health and wellbeing.* Stockholm, Reports from the Laboratory of Clinical Research, 1974 (in Swedish).

29  GHATA, J. La chronobiologie appliquée à l'hygiène de l'environnement. *Archives des maladies professionnelles, de médecine du travail et de sécurité sociale,* **32**: 385–393 (1971).

30  GHATA, J. ET AL. Rythmes circadiens désynchronisés du cycle social (17-hydroxy-corticostéroïdes, température rectale, veille-sommeil) chez deux sujets adultes sains. *Annales d'endocrinologie*, **30**: 245–260 (1969).

31  GOLDENBERG, M. Adrenal medullary function. *American journal of medicine*, **10**: 627–664 (1951).

32  GOSSLER, K. ET AL. Das Verhalten einiger biochemischer Parameter bei der physischen Beanspruchung durch einen Langstreckenmarsch. [The behaviour of some biochemical parameters in physical stress induced by a long-distance march.] *International archives of occupational and environmental health*, **41**: 103–115 (1978) (in German).

33  HADŽIOLOVA, I. ET AL. Changes in excretion biorhythm of some steroid hormones and catecholamines in urine during work. *Hygiena i zdraveopazvane*, **6**: 566–574 (1979) (in Bulgarian).

34  HALBERG, F. & NELSON, W. Some aspects of chronobiology relating to the optimization of shift work. In: *Shift work and health: a symposium.* Washington, DC, United States Government Printing Office, 1976, pp. 13–47 (DHEW Publication No. (NIOSH) 76–203).

35  HALBERG, F. & VISSCHER, M. B. Some physiologic effects of lighting. In: *Proceedings of the First International Photobiological Congress, Fourth International Light Congress, Amsterdam, 1964*, pp. 396–398.

36  HAYES, D. K. ET AL. Survival of the cradling moth, the pink bollworm and the tobacco budworm after 90° phase-shifts at varied, regular intervals throughout the life span. In: *Shift work and health: a symposium.* Washington, DC, United States Government Printing Office, 1976, pp. 48–50 (DHEW Publication No. (NIOSH) 76–203).

37  JENNER, D. A. ET AL. Catecholamine excretion rates and occupation. *Ergonomics*, **23**: 237–246 (1980).

38  JOHANSSON, G. *Activation, adjustment and sympathetic-adrenal-medullary activity. Field and laboratory studies of adults and children.* Reports from the Psychological Laboratories, University of Stockholm, Suppl. 21, 1973 pp. 1–25 (in Swedish).

39  JOHANSSON, G. Social psychological and neuroendocrine stress reactions in highly mechanized work. *Ergonomics*, **21**: 583–599 (1978).

40  KALIMO, R. ET AL. Psychological and biochemical strain in firemen's work. *Scandinavian journal of work, environment and health*, **6**: 179–187 (1980).

41 KRAHENBUHL, G. S. ET AL. Task-specific simulator pretraining and in-flight stress of student pilots. *Aviation, space and environmental medicine*, **49**: 1107–1110 (1978).

42 KUJALOVA, V. ET AL. Work strain evaluation by catecholamine excretion. In: Usdin, E. et al., ed. *Catecholamines and stress: recent advances.* Amsterdam, London, and New York, Elsevier, 1980, pp. 471–476 (Developments in Neuroscience Series, No. 8).

43 KVETŇANSKY, R. ET AL. Dopamine-$\beta$-hydroxylase in the plasma as an index of sympathetic activity in men.] *Vojenske zdravotnicke listy*, **44**: 22–25 (1975) (in Czech).

44 LEVI, L. Stress and distress in response to psychosocial stimuli. *Acta medica Scandinavica*, Suppl. 528, 1972, Vol. 191.

45 MAKAROV, V. I. Vlijanie fizičeskoj nagruzki na bioritmy čeloveka. [Effects of physical load on human biorhythms.] *Problemy kosmičeskoj biologii*, **34**: 130–135 (1977) (in Russian).

46 MANNINEN, O. Combined and single effects of prolonged noise and vibration exposure on employees' cochleovestibular functions and urinary catecholamines. In: Usdin, E. et al., ed. *Catecholamines and stress: recent advances.* Amsterdam, London, and New York, Elsevier, 1980, pp. 483–488 (Developments in Neuroscience Series, No. 8).

47 NEDELTCHEVA, K. Excretion of certain biogenous amines (free and conjugated) in urine during nervous-sensory and psycho-emotional work strain. *Agressologie*, **16**: 193–197 (1975).

48 NILSSON, K. O. ET AL. The influence of short term submaximal work on the plasma concentrations of catecholamines, pancreatic glucagon and growth hormone in man. *Acta endocrinologica*, **79**: 286–294 (1975).

49 PÁTKAI, P. ET AL. Field studies of shiftwork: I. Temporal patterns in psychophysiological activation in permanent night workers. *Ergonomics*, **20**: 611–619 (1977).

50 PÁTKAI, P. ET AL. The diurnal pattern of some physiological and psychological functions in permanent night workers and in men working on a two-shift (day and night) system. In: Colquhoun, P. et al., ed. *Experimental studies of shiftwork.* Opladen, Westdeutscher Verlag, 1975, pp. 131–141.

51 PIERCE, D. ET AL. Urinary epinephrine and norepinephrine levels in women athletes during training and competition. *European journal of applied physiology*, **36**: 1–6 (1976).

52 REINBERG, A. ET AL. Spectre thermique (rythmes de la température rectale) d'une femme adulte avant, pendant et après son isolement souterrain de trois mois. *Comptes rendus hebdomadaires de scéances de l'Academie des Sciences*, **262**: 782–785 (1966).

53 REYNOLDS, Y. Catecholamine excretion rates in relation to life-styles in the male population of Otmoor, Oxfordshire. *Annals of human biology*, **8**: 197–209 (1981).

54 RODAHL, K. & VOKAC, Z. Work stress in long-line bank fishing. *Scandinavian journal of work, environment and health*, **3**: 154–159 (1977).

55 RUTENFRANZ, J. ET AL. Desynchronization of different physiological functions during three weeks of experimental night shift with limited and unlimited sleep. In: Colquhoun, P. et al., ed. *Experimental studies of shiftwork.* Opladen, Westdeutscher Verlag, 1975, pp. 74–77.

56 SLOB, A. ET AL. The effect of acute noise exposure on the excretion of corticosteroids, adrenalin and noradrenalin in man. *International archives of occupational health*, **31**: 225–235 (1973).

57 SUDOH, A. Urinary excretion of catecholamines in various situations. *Bulletin of the Tokyo Medical and Dental University*, **18**: 295–310 (1971).

58 TIMIO, M. ET AL. Free adrenaline and noradrenaline excretion related to occupational stress. *British heart journal*, **42**: 471–474 (1979).

59 TORSVALL, L. ET AL. Age, sleep and irregular work hours. *Scandinavian journal of work, environment and health*, **7**: 196–203 (1981),

60 TOUITOU, Y. ET AL. Circadian rhythms in 11 urinary variables of 7 healthy young men. *Annales de biologie clinique*, **36**: 330 (1978).

61 TSANEVA, N. & DALEVA, M. Field study of the diurnal changes of the adrenal system. In: Colquhoun, P. et al., ed. *Experimental studies of shiftwork*. Opladen, Westdeutscher Verlag, 1975, pp. 206–212.

62 TSANEVA, N. ET AL. The effect of emotional stress on circadian rhythms of 17-ketosteroids, 11-oxycorticosteroids and catecholamines excretion in radiospeakers. *Acta medica Bulgarica*, **1**: 51–69 (1983).

63 VOKAC, Z. ET AL. Circadian rhythmicity of the urinary excretion of mercury, potassium and catecholamines in unconventional shift-work systems. *Scandinavian journal of work, environment and health*, **6**: 188–196 (1980).

64 WHITE, J. A. ET AL. Effect of physical fitness on the adrenocortical response to exercise stress. *Medicine and science in sports*, **8**: 113–118 (1976).

65 WINK, A. Lawaai en bijnierschors. [Noise and the adrenal cortex.] *Tijdschrift voor sociale geneeskunde*, **49**: 114 (1971) (in Dutch).

Chapter 7
# Mental disorders related to work
C. Dejours[1]

## Introduction

Whether work can so affect a man as to make him mad is a question that has long been asked, and one that can be raised technically in two ways.

1. *Are mental disorders of which work is the origin to be found only in certain situations—i.e., in relation to specific work?*

What is implied is that work is capable of giving rise to particular neuroses or psychoses. Such a supposition is based on a model copied from industrial toxicology which succeeds in establishing that a disease such as chronic lead poisoning results from the exposure of workers to lead vapour, or that silicosis results from exposure to silica dust.

In fact, none of the research carried out on the hypothesis that occupational mental disorders exist has yet yielded convincing results. This is not surprising, having regard to what is accepted in general psychopathology. Whether reliance is placed on psychodynamic and psychoanalytical models, in which the events of early childhood are seen as determinants in clinical forms of mental disorder, or on biological models, in which the etiology lies in neurochemical disorders often genetically determined, in psychopathology endogenous factors are most often held to be responsible for mental disorders, with exogenous factors, one of which is undoubtedly work, accorded no more than a subsidiary role.

2. *Does work contribute to the appearance of mental disorders that are not work specific, such as schizophrenia, hysteria, or depression, or at least to crises and acute attacks of such disorders, the symptomatology of which, once it becomes manifest, is independent of work?*

This is a question that has always been very difficult to answer, because all such disorders, when decompensated, and often even before decompensation becomes evident, are accompanied by a *deficiency syndrome*—i.e., an alteration in concentration, in endurance under strain, and in intellectual, cognitive, and productive performance (4). This amounts to saying that workers suffering from known mental disorders are almost always excluded from production, and that, once the

---

[1] Domaine du Grand Mesnil, Centre Hospitalier d'Orsay, Orsay, France.

disorder is established, it becomes virtually impossible to determine the part played by work from among the other factors—personal, familial, emotional, material, or congenital, long-standing or recent—involved in the etiopathogenesis of psychiatric decompensation. This has led some specialists to conclude that occupational psychopathology has no purpose.

This having been said, there exists an intermediate sphere, between the state of decompensated mental disease and that of *psychological wellbeing or comfort*, in which it may perhaps be possible to seek manifestations or disorders that, though disjointed, could be linked to work. The term *psychological distressed state* is applied to this sphere, lying between health and mental disorder. It is comparable to a common situation in somatic pathology. There is, for example, between the state of complete physical health and that of clinical chronic lead poisoning—with abdominal pain, headache, renal insufficiency, arterial hypertension, and neuropathy—a period during which the worker is not yet clinically ill; he is, nevertheless, already impregnated with lead, which is not a normal situation. This state of subclinical physical distress may be determined by such means as looking for stippled red cells or measuring urinary aminolaevulinic acid. Similarly, if the object is to demonstrate a state of psychological distress, recourse must be made to appropriate tools of investigation, which are not those used in classical psychiatry.

In this respect, studies based on the concept of stress provide an interesting approach (see Chapter 2), to the extent that they enable the identification of abnormalities that are not always obvious syndromes or disorders. Studies on the psychology of stress have been the first to relate work to the psychologically distressed state.

Such research does not yield specific or pathognomonic results. Whereas a specific task may produce specific modifications in the biological variables, it is not possible, in reverse, to take specific biological modifications and say that they stem from specific tasks.

In occupational psychopathology, research into the psychologically distressed state is seeking to define precisely the behavioural patterns, attitudes, and collective and individual defences that are specific to each type of work. Should this be achieved, the body of material assembled will enable the specific clinical features of the relationship between mental health and work to be described. It must be stressed that, at the present stage of research, efforts should concentrate on clinical studies not experiments reproducible in the laboratory, since to obtain the desired clinical picture subjective and qualitative investigations must draw on the experience of working people.

In order to analyse the psychologically distressed state, which may often be concealed or hidden, recourse could be made to sensitive tools capable of locating and revealing psychological phenomena that might otherwise remain unperceived. To that end the semiological methods and systematization suggested for the psychoanalysis of mental functioning might be brought into use (5). This would enable psychopathological reactions and the psychological defence systems customarily at work in individuals who are not mentally ill and who can successfully maintain their balance in the face of endogenous and exogenous influences to be taken into account.

## The relationship between mental health and work and the part played by the organization of labour

In studying the relationship between the individual and work, special attention is usually given to working conditions. Job characteristics—e.g., postural constraints—may be involved, or work surroundings, but matters relating to the organization of work, although acknowledged to be basic, often remain beyond the reach of change. This is due principally to the fact that it is technically, economically, and politically far more difficult in practice, in applying ergonomics for example, to influence the organization of labour than it is to alter working conditions. What should be understood by the term *working conditions* for the purposes of this discussion are physical working conditions (e.g., noise, vibration, temperature, radiation, lighting) chemical conditions (e.g., vapour, dust) and biological conditions (e.g., bacteria, viruses, yeasts). The term *organization of labour* means the *division of labour*, encompassing task content, mode of operation and work pace, and the correlative *deployment of individuals*—i.e., the way in which each worker, e.g., foreman, chargehand, typesetter, factory worker, or labourer, is assigned a place and a function in relation to others. In the same way as actual task content, working relations are thus regulated, delineated, and precisely laid down through the organization of labour.

The organization of labour conflicts with the psychological functioning of the individual at every point. The division of labour, more particularly task content, can raise questions in regard to motivation or, in more clinical terms, the desire to perform, the pleasure derived from, the "cathexis" of, a task; it can also give rise to dissonance because of the individual's history and personal wishes and the content of the work outsiders decide he should perform. The deployment of individuals can create a rigid framework within which interpersonal conflicts arise involving the emotions.

It can thus be understood that psychologically distressed states of some specificity may arise from the mode of organization of labour. Clinical investigations have effectively demonstrated that building workers do not experience the same type of psychological distress as workers in the processing industries, and assembly-line workers or office workers experience a different type again.

## Psychological distress in groups of workers forming teams: the defensive ideology

Although workers in other industries, especially the chemical industry, employ defences against anxiety, the building trade is used here for purposes of illustration. The numerous incapacitating and fatal accidents among building workers show that the risks are particularly great, yet their disregard of safety instructions is well known, as if they were oblivious to the risks or, on the contrary, courted them. Some authors maintain that the psychology of building workers is highly specific, characterized by a marked taste for danger and displays of physical prowess, pride and feelings of rivalry, manifestations of virility, and bravery; but also by recklessness to the extent of being oblivious to reality, lack of discipline, and individualistic tendencies. Disdain of the risk of accident is a well known attitude, as is the disregard of safety instructions. However, a disdainful attitude towards risk cannot be taken at face value as too often happens; it has a hidden and, as shall be seen, quite serious meaning. Attitudes of scorn, unawareness, and obliviousness are a facade and it cannot be conceded without question that the building workers themselves are the least aware of the risks they run. It has, in fact, been shown that the facade can collapse to reveal unexpected, staggering, and dramatic anxiety (1).

When the time for swaggering and defiance is past, a worker talks of the sites he has worked on and tells of the number of accidents he has witnessed, or of which he has been the victim. He speaks of friends killed at work and of the families of the injured and the disabled. He shows that he is more aware than anyone of the risk, and lives through it at the deepest level of his daily life. When he makes disclosures his tone of voice and the emotion he expresses leave no doubt as to his feelings in the mind of the listener. It is only necessary to have witnessed such an outburst to realize the intensity of the anxiety experienced in the face of danger, such as that caused by treacherous weather with the possibility of accident resulting in injury or death. Although these feelings exist, they surface only under exceptional circumstances and are mostly contained by the *defensive system*. The defensive system is absolutely necessary, since it is possible that were anxiety not to be neutralized in this manner but liable to break out uncontrollably during work, the worker would not be able to continue in the job. An acute awareness of the risk of falling, even without an abnormal emotional surcharge, would force a worker to take so many precautions that he would become inefficient. Some are prevented from continuing to work as a result of a proper assessment of the risk; this is not uncommon and fear, not always groundless, is an important cause of *occupational maladjustment* in the building industry. It appears, even outside work, in disguise in the form of the medical symptoms of anxiety—giddiness, headache, and various types of functional disability.

Denial and disregard are simple inversions and are inadequate strategies to ward off the anxieties of risk. Workers need more convincing evidence that they are overcoming those anxieties; that is how they come to add the further risk connected with personal achievement and contests of ability and bravery to that inherent in the work. In these contests they treat each other as rivals and in so doing they convince themselves that it is they who are creating the risk, rather than it being a danger to which they are exposed against their own wishes. To create, or worsen, a situation is to some extent to master it. This stratagem symbolizes for the workers the idea that they are retaining the initiative and keeping the upper hand over danger, and not vice versa.

The first distinguishing feature of this type of behaviour, the seeming unawareness of danger, results, in reality, from the defensive system, which controls anxiety. The second is its collective nature. It is characteristic of all who perform potentially dangerous tasks in the building trade. For the defensive system to be able to function in this way it has to find confirmation; only if all participate in the stratagem is its symbolic effectiveness ensured. No one must be afraid; no one must let it be seen that he is afraid; no one must stand aloof from the professional code; no one must refuse to make an individual contribution; there must never be talk of danger, risk, accident, or fear. These implicit rules are observed; the workers do not like to be reminded of what they seek to overcome at such cost. This is one of the reasons why safety campaigns encounter such resistance. All building workers are well aware that a safety harness does not always prevent an accident. To force them to wear one is to remind them that the danger really exists, making the job even more difficult to perform because of the greater burden of anxiety. Consequently, the resistance to safety measures encountered in the building trade does not stem from apparent recklessness or immaturity; it is a deliberate way of behaving, the object of which is to make more tolerable a risk that would hardly be lessened by scant safety measures.

It can be seen that, to be effective, the defensive system requires great cohesion and solidarity in the face of death. This is undoubtedly the reason why certain behaviours become a tradition in the trade, or even a true *defensive ideology*, characteristic of the occupation. An ideology needs its sacrifices and its martyrs. Accidents definitely result from this dangerous way of behaving and the competitive feats of daring. However, what is made possible by these sacrifices should be examined. The colleagues of a worker killed on the job may say "He was killed because that was what he wanted; death was what he was seeking; he went too far". It may be true that he was courting disaster, but what has been made possible is a perpetuation of the idea that all that is necessary not be a victim is not to wish to be one, a formula highly capable of calming anxiety. Without such an analysis, it is not possible to understand why workers who are aware of the rashness of their actions can so often assert that an accident was the fault of the person concerned.

The third distinguishing feature of this type of behaviour is its functional value *vis-à-vis* productivity. It is functional for the workers on the building site who share it, but in another way it is functional for those who are not directly involved. Should a worker not be able to accept the defensive ideology for himself, should he not be able to overcome his fear, he will have to stop working. The group, defended by the ideology, eliminates the individual who is not able to face up to the risk. That is why the most vulnerable amongst a group of workers becomes a target of scorn for the others. If he cannot abandon his fear he will sooner or later be eliminated—i.e., rejected. In rejecting him not only has the group operated a real process of selection, which is a guarantee of the operational worth of every worker remaining on the site, but it has defended itself against the anxiety aroused in each individual, and collectively in the group as a whole, by the remarks and behaviour of the "coward". It may, therefore, be conceded that the defensive ideology plays a very important role in work continuity.

Another example relates to the so-called initiation of young workers newly arrived on the site. It is not uncommon for them to be cruelly put to the test. They are ribbed during meal breaks on their manliness, required to perform physical feats, watched, and made to run the gauntlet of the defensive ideology. If they emerge victorious, they become full members of the group, assuming the constituent elements of collective defence. If they are not able to tolerate this climate they have to resign, which happens from time to time. The defensive ideology is, therefore, functional as it concerns the group, enabling it to express its cohesion and courage, and also as it concerns work, since it guarantees productivity.

When the behaviour of the workers is interpreted in this manner their apparent obliviousness takes on new meaning; what it takes to achieve it is the price they have to pay to cope with the burden of anxiety inherent in the work. The part played by wine and spirits may be examined in relation to this ideology. They are the burst of energy, not so much physical as psychological, that helps the worker to cope in the type of organization of labour that is dangerous. Before working, a drink is of assistance because of its symbolic value and its anxiolytic psychopharmacological activity. It is not by chance that the psychological role of drink fits both the traditions and the living habits of the workers. Furthermore, it is in harmony with the thirst produced by physical exertion.

So the anxiety aroused by the risks of the job is not always evident in what the workers say. It has to be looked for behind the defensive attitudes. There are many occupations in which defensive systems deeply structured by the nature of the risk involved are to be found. In comparing some occupations there are analogies between the systems, but in others the systems are completely different and specific to the occupation. Such is the case in relation to the chemical industry, in which a defensive ideology exists but possesses quite different characteristics from that of the building trade.

Lastly, there is another characteristic of the defensive ideology that should be emphasized. In order to arise it requires the existence of a group of workers—i.e., not merely a number of workers employed in the same place but a number constituting a team among whom the tasks of a job are divided. When work is broken down into separate, repetitive tasks, with very little communication between the workers and extremely rigid organization of labour, there is little place for the development of a defensive ideology, and only traces will be found.

## Implications in the field of general psychopathology: the subjective post-traumatic syndrome

It is useful to study the repercussions of distress at work on the sequelae of occupational accidents. A condition very frequently encountered as a psychiatric complication of an occupational accident is the *subjective post-traumatic syndrome*. It appears in general after a wound has healed, a fracture has knitted together, or acute poisoning has been cured. It is characterized by a wide variety of functional disorders—i.e., disorders without organic substrates—or by the abnormal persistence of a symptom that became apparent after the accident. Thus, though a scalp wound caused by a falling stone is stitched and heals within two weeks, the injured person continues to complain of such symptoms as headache, tingling of the surface of the skull, or strange sensations in the head. Clinical and paraclinical investigations fail to reveal anything, but the subjective symptoms often prevent the patient from resuming work. What can then follow is a dialogue of the deaf between the worker, the physician, and the social security authorities, leading, after a few months, to a state of chronic demand in the patient.

The usual interpretation given to this syndrome is that it is the hypochondriacal compensation of an underlying neurotic structure pre-dating the accident. The accident itself is merely a *reactional factor* or a triggering element. It is possible, however, to interpret this syndrome differently, as stemming from work and not from the worker's mental structure. The main effect of the subjective post-traumatic syndrome is to prohibit the resumption of work, or to compel an occupational reclassification to a trade without physical risk; this is the key to understanding it. It is as if the accident reveals the ineffectiveness of the defensive ideology of the occupation as a protection against danger. The victim of the accident retains an acute and conscious perception of the threat present in the work. Since he would be incapable of concealing a justified anxiety, the conscious or unconscious aim of the worker is not to return to work so that he no longer has to face up to the risk. This aim could equally well be expressed in different terms: as he would, henceforward, be incapable of entering with his colleagues into a defensive system in which he no longer believes the worker must, if he returns to work, carry on with the job with a true awareness of the danger. This is precisely what is impossible for a normal individual. Moreover, he is unable to express this anxiety, since to do so might be equivalent in the minds of many—work-mates, physician, family—to a loss of manhood, or to simple refusal to work. In fact, it is not the work but the risk that is refused. At best a change of work is obtained. In other cases, however, to remain ill is the only admissible excuse for not being able to return to work. This strange disorder, the subjective post-traumatic syndrome, is extremely common, affecting several thousand workers involved in accidents every year. Consequently, the hypothesis that it is essentially determined by work and not by a neurotic personality structure can be advanced for two reasons:

(a) In the psychosomatic study of patients suffering from the post-traumatic syndrome it is shown that a very wide variety of psychological structures is to be encountered; there is no preformed neurotic characteristic of the syndrome;

(b) Although both physicians and psychiatrists are accustomed to attribute the occurrence of the syndrome to a hypothetical pre-formed neurotic structure, unlike some other spheres of neurotic psychopathology it exhibits exceptional resistance to psychiatric treatment. Above all it cannot be deciphered by psychotherapy, unlike other neurotic symptoms; as far as can be ascertained there has never been a description in te literature of a successful analysis of the syndrome through psychotherapy. That it is not amenable to analysis must stem from the fact that it is determined by sociooccupational and not psychoaffective factors. Its meaning and significance are not to be found in the patient's past history; on the contrary, they are present in the nature of his working conditions and the organization of labour.

## Psychological distress in unskilled occupations

A very different working situation is that in which the psychologically distressed state assumes characteristic forms.

The extreme division of labour, as on certain assembly lines, raises the question of *monotony*. Is it possible to foresee the psychological consequences of monotonous work?

The questions raised when considering the problem of monotony are fairly complicated. How is it that a normal individual is able to tolerate mentally an operative cycle lasting at most a few seconds, repeated for hours, months, years, or an entire working life-time? Such conditions are completely abnormal, not spontaneously encountered in the natural environment. On the contrary, recent advances in biology and psychology demonstrate that stability is not the rule in the living world. What have been referred to as *constants* are in fact variables, and life is a sequence of random movements in which there is ceaseless interplay between balance and imbalance. The conditions resulting from repetitive work imposed on man by the organization of labour oppose those biological and psychological movements, which they sorely try, and appreciable mental difficulties result.

Assembly-line work might be considered as the classic example, but there is work consisting of less fragmented repetitive tasks that raises the question of how a worker succeeds in resisting the monotony caused by not knowing why he is performing a task and not understanding its purpose. In psychopathological terms, the task can be said to have no meaning in relation to the worker's history; his past, his wishes, his childhood aspirations, or his fantasies.

Direct psychological *cathexis* in relation to the content of the task is not impossible. The worker is able to tolerate the situation only through a lateral cathexis—his wages—which should perhaps be termed "the wages of suffering", the reward for facing up psychologically to monotony or absurdity. The task cannot, therefore, be fitted into a longitudinal or diachronic perspective; its history ceases to exist on the production line or the machine tool. Furthermore, sublimation is ruled out by the mode of operating and the monotony, which can have no personal significance because they are controlled by someone else. If there is no direct cathexis, no sublimation, what is to become of the impulses and wishes of the individual? What place is there in this peculiar situation for fantasizing (3)?

At best, fantasizing enables the worker to escape from chronic frustration and mental suffering. However, the organization of labour can be such that the task itself is in opposition to fantasizing activity. Clinical studies have shown that evasion through fantasizing can be difficult to achieve because the fantasizing is so mutilated by work cycles that it becomes ineffective and sometimes cannot be brought into play; or because it is quite simply impossible, as when piece work is involved—i.e., repetitive work under pressure—in which the wages needed to make up an adequate minimum income are dependent on the number of items produced. In that case, the tempo of work is so great that antagonism appears between fantasy and work; fantasizing must reduce attention and concentration, and consequently output, since it suggests other things, other action, a different existence; but it can also suggest clumsy action, faulty operation, an accident at work, or an increased number of defective products which must then be made good.

However, this theory that there is antagonism between repetitive work and psychological functioning (fantasizing activity in particular) is open to dispute, for it is not easy to make psychological functioning cease; except in mental patients such functioning is never spontaneously interrupted. In order to put a stop to fantasies and psychological functioning an employee in a system with an extreme division of labour would have to struggle against the most precious part of his personality.

## The psychopathological consequences of monotonous work

Although the theory that there is antagonism between monotonous work and spontaneous psychological functioning may be open to dispute, the fact remains that when changing work cycles are involved fantasy may be ruled out. However, the repression of psychological functioning is not a simple matter; it calls for constant effort on the part of the worker and this employs a considerable amount of energy. The result is fatigue; it is not at all surprising that monotonous or uninteresting tasks exhaust the worker, even though the physical load is not excessive. It is not the work itself that exhausts him but the struggle forced on him by the organization of labour, against the most vital part of his personality. Thus the sequence— *fatigue–asthenia–depression*—is already implicit. In addition the repression of psychological functioning for several hours a day may not be immediately reversible. There are many observations to show that arrested spontaneous functioning does not resume involuntarily and instantaneously. Often several days are needed before a worker, freed from the constraints of work, rediscovers desires, tastes, longings, and designs. This difficulty in recovering increases with age. The worker himself is painfully aware of this distortion or alteration, and this contributes to narcissistic collapse. To feel that he is becoming stupid and to experience psychological impoverishment is very painful to him and contributes to depression.

The worker often finds it so difficult to arrive at the state of repression of psychological functioning that once having achieved it he attempts to maintain it away from work by various devices, ranging from the pursuance of frenetic activity unjustified outside work, to ensure that the conditioning continues, to passivity—e.g., in front of the television—to avoid the emergence of spontaneous mental functioning that would have to be combatted the following morning. The worker himself involuntarily contrives his own conditioning in order to suffer less. Again, he is often only too aware of this absurd logic, and this contributes to feelings of humiliation and shame, which are also at the heart of depression.

The state encountered when mental functioning is paralysed is precisely what experts in psychosomatic disease describe as *essential depression* (7). This type of depression, which is readily distinguishable from the depression of mourning or melancholy, is often difficult to diagnose clinically for anyone not familiar with the semiology of mental functioning, because of the paucity of characteristic symptoms. It comprises characteristic traits described as operative thought (8) or alexithymia (9). Essential depression is always diagnosed as a psychopathological state preceding, and very often accompanying, the onset of a serious somatic illness or the appearance of a complication to a chronic condition (6).

The final outcome of the singular relationship between unskilled, monotonous work and depression may often be somatic, not mental. For precisely that reason the relationship becomes indecipherable. It could also be that the methods of command and surveillance of those engaged in unskilled tasks place the individual under strain, submitting him to mental suffering and humiliation that strengthens and echoes the spiral of depression.

## Managerial workers and skilled workers

What has been described in relation to unskilled tasks cannot apply to the type of depression experienced by managerial or skilled workers, who often come to mind when reference is made to occupational depression. Their work is complex, and they usually have a free choice of tasks and strong motivation, which generally presupposes fairly long training and, above all, freedom to intervene in the day-to-day organization of their work. Their situation cannot be compared to that of unqualified employees.

Nevertheless, the general trend in society today is towards an increasing division of labour. From the psychological point of view this signifies that the work content becomes specialized and increasingly fragmented, the craft skill and knowledge progressively lower, and the standardization of working procedures and control by various devices—e.g., supervision, administration, computerization—more apparent. The worker begins to feel deprived of freedom of action and the liberty to intervene in the organization of his work. Managerial staff are caught up in this logic, of which they are sometimes painfully aware. This deprivation and profound transformation in their psychological relationship to work are beginning to affect professions that have hitherto felt themselves to be immune from this particular repercussion of industrialization. Certainly it is demonstrated daily in clinical practice that tiredness, disappointment, deception, and sometimes ill-boding resignation or, on the contrary, disturbing anger, are apparent in the conversation of patients who are not always being treated for depression.

## Conclusions

The clinical conditions described in connection with the two working situations considered in this chapter—work in the building industry and repetitive work—provide examples of the knowledge that can be gained from studies in occupational psychopathology. It has been shown that, even if work is not responsible for mental disorders, the organization of labour creates specific disorders and distress that cannot be considered anecdotal or harmless. By fostering distress and frustration it affects the entire mental functioning of the worker, on whom it imposes constraints that are sometimes severe. The worker is able to resist the chronic distress imposed on him only by summoning defence mechanisms that use up a great deal of his energy and produce a fatigue that becomes an integral part of his work-load. Together with the physical and nervous work-loads it is legitimate to accord a place to the psychological work-load (2). The defence mechanism leads the worker into unusual or paradoxical behaviour and attitudes which, thanks to advances in occupational psychopathology, are now understood. This type of distress weakens the worker and modifies his health; when the balance is disturbed it is not always mental disorder that arises. For example, decompensation, when it results in that particular form of depression called essential depression, gives rise to somatic illness. It has been postulated that the psychological stress resulting from certain forms of the organization of labour helps to shorten the lives of the workers involved. Apart from revealing the mechanisms of occupational psychological distress, occupational psychopathology clarifies fundamental questions concerned with health and prevention, demonstrating that work organization must be changed in order to reduce psychological distress.

## References

1 DEJOURS, C. *Travail: usure mentale.* Paris, Editions du Centurion, 1980.

2 DEJOURS, C. La charge psychique du travail. In: *Equilibre ou fatigue au travail?* Paris, Les Editions ESF (Actes des journées de la Société Française de Psychologie: Section Psychologie du Travail), 1980, pp. 45–55.

3 DEJOURS, C. Désir ou motivation? L'interrogation psychanalytique sur le travail. In: *Quelles motivations au travail?* Paris, Les Editions ESF (Actes des journées de la Société Française de Psychologie: Section Psychologie du Travail), 1982, pp. 118–126.

4 EY, H. *Des idées de Jackson à un modèle organo-dynamique en psychiatrie.* Toulouse, Privat, 1972.

5 FREUD, S. The ego and the id (1923). *The standard edition of the complete works of Sigmund Freud.* London, The Hogarth Press, 1961, Vol. XIX, pp. 1–66.

6 MARTY, P. A major process of somatization: the progressive disorganization. *International journal of psychoanalysis,* **49**: 246–249 (1968).

7 MARTY, P. La dépression essentielle. *Revue française de psychanalyse,* **32**: 595–598 (1968).

8 MARTY, P. ET AL. *L'investigation psychosomatique.* Paris, Presses Universitaires de France, 1963.

9 SIFNEOS, P. E. The prevalence of "alexithymic" characteristics in psychosomatic patients. *Psychotherapy and psychosomatics,* **22**: 255–262 (1973).

Chapter 8

# Psychosomatic disease as a consequence of occupational stress

Lennart Levi[1]

## Introduction

The causal relationship between exposure to occupational stressors and psychosomatic and psychiatric morbidity has many links that are complex, interrelated, non-linear, and conditioned by a multitude of influences, at work and outside it, often over a long period. There is a saying that it takes three generations to make a gentleman; the developments leading to arterial hypertension, myocardial infarction, or suicide, may similarly span a long period.

In this chapter problems affecting the health and wellbeing of workers in different countries and their possible relation to psychosocial factors in working life will be considered.

In Sweden, according to information published by the National Board of Health and Welfare (41):

— Every third adult suffers from malaise, sleep disorders, fatigue, dejection, or anxiety.
— Every seventh working person is mentally exhausted at the end of the working day.
— Every other man and three women out of four will suffer from pronounced mental decompensation (breakdown) on one or more occasions between birth and the age of 60 years.
— Every tenth man has an alcohol-related problem.
— In a population of 8.3 million, 2000 people commit suicide each year; 20 000 attempt to do so.

Taken together these figures mean that approximately every third or fourth Swede lives a life in which malaise, anxiety, fatigue, or dejection are common components.

In a report of the Commission on Mental Health to the President of the United States of America mental health problems in the USA are closely related to psychosocial environmental factors, such as unrelenting poverty, unemployment, and race, sex, and age discrimination (43). For example, persistent, handicapping mental health problems occur in 5–15% of children. Alcohol abuse is a major social, physical, and mental health problem, with an estimated annual cost to the nation of over US$ 40 000 million. Nearly 15% of the population needs some form of mental health service at any one time, and it

[1] WHO Psychosocial Centre, Laboratory for Clinical Stress Research, Karolinska Institute, Stockholm, Sweden.

is estimated that up to 25% suffer from mild to moderate depression, anxiety, and other symptoms of emotional disorder.

Reports from a number of countries in Europe give a similar picture (8, 11, 21, 32). Reviewing an impressive body of evidence, the consensus is that:

— The general level of physical and psychological strain in the working population is high.
— There appears to have been an increase, during the 1970s, in the amount of strain experienced.
— There are marked differences in the types of strain experienced between individuals of different socioeconomic status.
— The number of complaints of a mental nature among younger people is increasing.

It has been claimed that about half the entire working population are unhappy in their jobs and as many as 90% may be spending much of their time and energy in work that brings them no closer to their goals in life. About 75% of those who consult psychiatrists are experiencing problems that can be traced to a lack of job satisfaction or an inability to relax (7, 8, 12, 17, 21, 26, 47).

## Mechanisms linking stress at work to disease

The possible mechanisms linking joyless devotion to hard work to stress and disease pertain to cognitive, emotional, behavioural, and physiological reactions (26, 27).

The cognitive reactions to exposure to psychosocial occupational stressors include restrictions of the scope of perception and the ability to concentrate that are needed to be creative or to make adequate decisions. Although it is difficult to assess the amount of ill health induced through such reactions, it seems highly likely that they do, indeed, contribute not only to frustration and low job satisfaction but also to accidents at work.

Emotional reaction, such as anxiety or depression, in response to a great variety of environmental stressors is part of everyday experience, as is a hypochondriac reaction—i.e., when the subject becomes exceedingly and unpleasantly aware of the normal neural feedback to the brain from the various parts of the body. As this last state depends on the worker's interpretation of the normal signals generated in the body, neither organ function nor organ structure is disturbed. Persons experiencing it tend to go from doctor to doctor undergoing many diagnostic tests, often without getting a satisfactory explanation of the true nature of their complaints or obtaining relief from them.

Similarly, it is well known from everyday experience that psychosocial stimuli influence human behaviour in a potentially disease-provoking manner. The use and abuse of alcohol, psychoactive drugs, and nicotine have been seen as results of sociocultural pressure—e.g., as means of achieving acceptance or as a reaction to increased responsibility—or as self-treatment for mental or physical distress. High suicide rates have been found in areas characterized by overcrowding or high population mobility, and in a large proportion of persons living alone.

It is not possible to list here all, or even most, of the physiological reactions that are assumed to constitute the possible pathogenic mechanisms. Although there are probably many others, those in the two following broad categories are considered to be particularly relevant and have been the most studied: neuroendocrine reactions, involving the hypothalamo-adrenomedullary and hypophyseo-adrenocortical axes, with secondary effects on lipid, carbohydrate, protein, and mineral metabolism; and reactions in the immune system (10).

## Work-related morbidity and mortality in psychosomatic diseases

The relationships between working environment, organization of labour, and work content, on the one hand, and pathogenic mechanisms, morbidity, and mortality on the other, have been studied by integrating the concepts and methodology of psychophysiology and epidemiology. In this way job dissatisfaction and physiological stress reactions can be associated with various job characteristics and with general and specific mental and psychosomatic ill health (28). Research designs have included three logically consecutive steps:

(1) Problem identification through surveys and the collection of morbidity data;

(2) Longitudinal, multidisciplinary studies of the interaction of high-risk situations and high-risk groups compared with controls;

(3) Evaluation of controlled intervention, through both applying the results of laboratory experiments and appropriate therapeutic or preventive interventions in real life.

A growing body of indirect evidence, which is reviewed below, suggests that causal relationships do exist and may be of considerable importance in the etiology of some major occupational health problems.

## Studies on psychosomatic morbidity in response to psychosocial stressors during work and leisure

Studies on the subject are either experimental, epidemiological, or both. In the experimental studies, animals or human volunteers are subjected to a variety of work-related stressors, and subsequent functional and (in the animals only) structural changes are demonstrated. In the epidemiological studies, groups of workers are studied retrospectively, cross-sectionally, or prospectively with regard to occupational exposure and subsequent morbidity and/or mortality. In the combined studies, environmental modifications (improvements) are introduced and evaluated.

In a review of an impressive body of animal studies, it was concluded that acute psychosocial stimulation can lead to the death of an animal when it is made helpless or cannot escape from an aggressor's threats (15). The collapse may be due to acute heart failure or kidney failure. Chronic stimulation can also lead to death, arteriosclerosis of the large vessels as well as of the heart and other organs being a cause of the fatal outcome. An important factor in producing lesions is the animal's perception of being powerless to induce the changes desired. Colonies of mice in sustained conflict over territory develop a progressive increase in blood pressure and heart weight, together with arteriosclerosis throughout the vascular bed. About 6 months of competition in a standard-population cage will induce permanent arteriosclerotic deterioration. These "middle-aged" mice develop classic arteriosclerosis; irreversible enlargement of the adrenal cortex and progressive renal damage, with death in uraemia, also occur.

Although animal experiments provide evidence of causal relationships they must be interpreted with caution, because of the artificiality of many experimental settings and the undeniable difference between mice and men. Most experiments with human subjects—in laboratory settings and in real life (26, 27, 28)—provide data relating to the influence of psychosocial occupational stressors on potentially disease-provoking mechanisms, but not to morbidity and mortality, since disease in human subjects can never be induced intentionally in order to test a hypothesis. A way out of this dilemma would be to carry

out "anti-stress", instead of stress, experiments—i.e., to eliminate naturally occurring exposures to occupational stressors in
one group but not in another over the same period of time, and
to study the reactions when (a) both groups remain exposed to
the naturally occurring stressors and (b) when the naturally
occurring stressors have been eliminated in one group (25). So
far, however, there have been few such combined experimen-
tal/epidemiological studies (2, 5, 28).

The more traditional epidemiological studies, whether retrospective or prospective, provide associations, often correlations,
between occupational exposure and subsequent ill health—i.e.,
circumstantial evidence. But association is not causation—
greying hair is a sign of approaching old age and an eventual
increased risk of mortality, but it does not cause those effects
and they cannot be prevented by dyeing the hair. One way of
enhancing such circumstantial evidence is to study the lapses of
time between a change and an adverse effect (3, 13).

Brenner started by examining whether or not illness follows a
period of exposure to life changes or stresses (3). Using
multivariate time series methods on readily available national
statistical environmental and health data, he formulated a
predictive equation that included the unemployment rate, the
inflation rate, real per capita personal income, and demographic
variables relating to the rates for each of several types of
pathological reaction—e.g., first admission to a mental hospital,
total mortality, suicide, homicide, cardiovascular and renal
disease mortality, death in cirrhosis of the liver, and imprisonment. He found that, in all cases, the national unemployment
rate was associated significantly and positively with increases in
the rates for each type of pathological reaction over a 5-year
period, starting with the year after the effect of the increased
unemployment rate was felt. He assumed that reactions of an
acute type—e.g., admittance to a mental hospital, imprisonment,
suicide, homicide—would predominate during the first 3 years,
and the more chronic reactions—e.g., cardiovascular and renal
disease, cirrhosis of the liver, and death through disease—during
the second 3 years. In general, his assumptions were correct.
Brenner further called attention to two often-overlooked
principles: (i) the *principle of acceleration*—deleterious life
changes are capable of producing stresses which in turn lead to
other life changes and stresses, e.g., loss of job may lead to
financial disruption, marital and parent–child strains, and
possibly breakup of the family; (ii) the *principle of contagion*—the
multiplier effect of one person's stresses upon another's.

Brenner's are all aggregate studies. In contrast, Rahe and others (34, 35, 36) based their analyses on data for individuals. In both retrospective and prospective studies they showed that the greater the number and intensity of changes in a specific subject's life over a certain period the more he is at risk of a subsequent decrease in health status. It may well be that such life changes, i.e., to a large extent changes in social structures and processes, confront the human organism with the necessity to adapt and it reacts with phylogenetically old non-specific preparation, i.e., stress in Selye's sense, facilitating fight or flight, increasing the rate of wear and tear on the organism, and eventually leading to the risk of morbidity and mortality. Such an assumption is supported by the results of a longitudinal study lasting 2–4 months in 21 male patients who had recovered from myocardial infarction (42). A positive, significant correlation was demonstrated between individual weekly mean life-change scores and an index of "stress" defined by Selye—the 24-hour level of catecholamine excretion—on the penultimate day of the same week.

A high life-change score often represents a situation characterized by overstimulation, which can be quantitative or qualitative, or both. Quantitative overstimulation, e.g., work overload, can also coincide with qualitative understimulation, e.g., tasks that are too simple, repetitive, or monotonous. Other predominantly situational occupational stressors believed to be of pathogenic significance include a conflicting or ambiguous role (e.g., that of a supervisor loyal towards both the management and the workers) lack of opportunity to exert control over one's own situation at work and elsewhere (low decision latitude) and lack of support from fellow workers, the boss, friends, and family members. Such conditions can interact, not only with one another but with such characteristics of an individual worker as genetic vulnerability; a propensity for being aggressive, competitive, or work-oriented, with a constant sense of urgency (Type A behaviour); a propensity for having unrealistic goals, low self-esteem, incapacity to unwind, inability to accept what cannot be changed, and a pathogenic life-style— e.g., in relation to alcohol, tobacco, food, exercise, and health care; a propensity for reacting by "giving-up". Buffering factors are believed to include an ability to cope effectively and having access to compensatory activities and experiences outside work.

Evidence has been presented in support of each of these factors, usually in an attempt to isolate a single "cause" of work-related psychosomatic morbidity or a series of linear causal factors. However, most probably a complex, interactional, non-linear systems model is needed in order to study and analyse them.

## Contribution of psychosocial factors to psychosomatic disease

As has been described above, psychosocial factors can influence health and caring for health, at work and elsewhere, through four mechanisms—cognitive, emotional, behavioural, and physiological. It follows that psychosocial factors contribute to literally every state of disease. Sometimes their contribution is marginal; sometimes it is of decisive importance. For obvious reasons it is not possible, in the present context, to review the role of psychosocial occupational factors in every possible type of disease. Detailed reviews have been carried out by such authors as Elliott & Eisdorfer (*10*), Henry & Stephens (*15*), Kahn (*17*), and Levi (*23, 26*).

Disturbances in bodily functions commonly found in workers exposed to stressful situations in working life include (*30, 44*):

— *muscular* symptoms—e.g., tension and pain;
— *gastrointestinal* symptoms—e.g., dyspepsia, indigestion, vomiting, heartburn, constipation, and irritation of the colon;
— *cardiac* symptoms—e.g., palpitation, arrhythmias, and inframamillary pain;
— *respiratory* symptoms—e.g., dyspnoea and hyperventilation;
— *central nervous system* symptoms—e.g., neurotic reactions, insomnia, weakness, faintness, and some headaches;
— *genital* symptoms—e.g., dysmenorrhoea, frigidity, and impotence.

Although such everyday symptoms may often be considered trivial by the physician, they can cause much distress and suffering for the patient, high cost to the community, and very considerable losses for the employer. Cardiovascular symptoms have attracted the most attention among the so-called *psychosomatic disorders*, particularly coronary heart disease and essential hypertension.

### Coronary heart disease

There is no doubt whatever that acute stressors can precipitate angina pectoris, arrhythmias, congestive heart failure, stroke, myocardial infarction, or sudden cardiac death in those who already have cardiovascular disease. In contrast, the part played by chronic influences is still open to question. It is said that they could affect the pathogenesis of coronary heart disease in predisposed healthy people, but both the etiology and the pathogenesis of coronary heart disease are multifactorial and knowledge in regard to them is incomplete (37).

In about 50% of cases the development of coronary heart disease can be explained by the presence of what may be called a conventional risk indicator, such as hypertension, hyper-lipoproteinaemia, or smoking. The remaining incidence is still to be explained. It seems likely, but is still to be proved, that psychosocial factors play an important role, in their own right and by influencing the conventional risk factors.

An analysis, by occupation, of arteriosclerotic heart disease mortality among white males aged 20–64 years in the USA showed that such occupational groups as college presidents, professors, instructors, and teachers have lower than expected rates, whereas lawyers, judges, physicians, surgeons, pharmacists, and insurance and real estate agents show the reverse pattern (19, 40). Since there was a two-to-one difference in rates, though such factors as social status and level of physical activity were similar for the members of all the groups, serious scrutiny of the respective work situations was considered to be justified. This was supported by separate findings (16) indicating that:

(1) Persons with coronary heart disease are significantly more dissatisfied with their jobs than others, and cardiac risk factors are related to decreased self-esteem;

(2) Increases in objective and subjective measures of job pressure, as reflected in high levels of work-load and responsibility, role conflict, and role ambiguity, are generally associated with coronary heart disease risk factors;

(3) Occupational mobility or frequent changes within an occupation predispose an individual to coronary heart disease.

Other studies bring into dispute the common assumption that coronary heart disease is a manager's disease. It is, in fact, more common in the lower social groups, among those with little education and low decision latitude but with a heavy work-load in a joyless job (18).

A positive correlation between a more easy going lifestyle and a low risk of coronary heart disease was demonstrated among the descendents of Italian immigrants who went to Roseto, Pennsylvania, USA in the late 1890s and who differ from the population of neighbouring communities in three respects (45). They live a quieter and more contented life; they have a considerably lower lipid content in their blood even though they eat at least as much fat as their neighbours; and their coronary heart disease mortality rate is less than half that of the neighbouring, more competitive, communities and, moreover, of the national average (9, 10, 31, 46).

It seems likely that, jointly with factors that limit vascular oxygen supply, centrally controlled neuroendocrine mechanisms dominate myocardial pathogenesis (33). By interfering with myocardial oxygen economy (catecholamines) and carbohydrate metabolism (glucocorticoids), those mechanisms influence the vital myocardial electrolyte equilibrium (loss of potassium and magnesium, gain in sodium) thus disturbing stimulus formation and conduction as well as cell contractility and structure. In a study in animals as disease models, using electrolyte steroid cardiopathy with necrosis (ESCN), it was found that degenerative myocardial changes may be provoked by stress, corticoids, catecholamines, electrolytes, or lipids. (39). It was pointed out that those agents make up a virtually complete list of the potential pathogens that have been given serious consideration by clinical investigators. It was possible to produce electrocardiogram changes, with ST-T depression, and eventual structural changes in the myocardial tissue, in the leader of a baboon pack simply by "dethroning" him (20). After a stressful 75-hour vigil, about 1 out of 4 in a group of 63 healthy military subjects developed reversible electrocardiogram changes, notably ST-T depressions, most conspicuously in lead $CR_5$ (22, 24).

With respect to sudden cardiac death, a review was published in 1971 of the evidence supporting the thesis that fatal cardiac arrhythmias, with or without associated myocardial infarction, may often be attributable to undamped autonomic discharges in response to either afferent information from below, or to impulses resulting from integrative process in the brain involved in adaptation to life experience, or both. Regulatory inhibition appears to be diminished in situations that are interpreted as overwhelming and without hope, such as total social exclusion and other circumstances characterized by hopeless dejection or sudden fear. At such times, the loss of regulatory inhibition may provide a mechanism of death. On the other hand, it would appear that the enhanced lability of autonomic responses associated with weary dissatisfaction, frustration, the feeling of abandonment, and dejection may be damped by emotional support from people and circumstances in the environment (45).

## Hypertension

Man's haemodynamic reaction to a variety of psychosocial stimuli comprises a visceral vasoconstriction and a muscular vasodilation, which prepare the body for muscular activity (4). Today, this type of reaction is generally not dissipated in muscular exercise, as it tended to be 20 or 30 generations ago. The blood pressure rises but usually rapidly returns to normal. In some subjects, however, especially those with a certain personality structure or coming from a hypertensive family, the reaction is not only exaggerated but tends to be protracted. It has been hypothesized that if this type of reaction occurs frequently, it may, by causing some of the links in the chain of physiological events to be overtaxed, eventually become dys-regulated, protracted, and fused.

It appears that certain occupations in which there is frequent exposure to mental stress, an overload of responsibility, or often occurring conflict, are more frequently associated with hyperten-sion. Thus an 11.8–74% incidence of hypertension was found among teachers and bank clerks compared with a 0.8–4.2% incidence among miners and labourers (38). A high incidence of essential hypertension was found among telephone operators employed at a large exchange, whose work entailed permanent mental stress without a moment's respite (29). In a similar way is explained a higher incidence of hypertension among engine drivers than among stokers (14), and among taxi drivers (1). The results of a study of the incidence of hypertension, peptic ulcer, and diabetes in air traffic controllers and second-class airmen support those findings (6). The data were obtained from annual medical examination records. The air traffic controllers were found to have a higher risk of developing hypertension than the second-class airmen, the added risk being related to working at high traffic density towers and centres. The air traffic controllers were also found to have a higher incidence of peptic ulcer and, to a lesser extent, diabetes.

It is not known whether a repeated transient hypertensive state leads to permanent hypertension. Visceral conditioning has been mentioned as one of many possible mechanisms. In a review and discussion of an impressive number of epidemiological studies it was concluded that a man, living in a stable society and well equipped by his cultural background to deal with the familiar world around him, will not show a rise in blood pressure with age (*15*). However, when radical cultural changes disrupt the familiar environment, with a new set of demands to which he has not been accultured, his social assets become of critical importance. Should they fail to protect him, he will be exposed to emotional upheavals and ensuing neuroendocrine disturbances that could result in cardiovascular disease.

# References

1 ALEKSANDROW, D. Studies on the epidemiology of hypertension in Poland. In: Stamler, J. et al., ed. *The epidemiology of hypertension: proceedings of an international symposium.* New York and London, Grune and Stratton, 1967, pp. 82–97.

2 ANDERSSON, L. & ARNETZ, B. Psychological, medical and social measures of increasing the activity in old age as a means of prevention of social isolation and reduction of needs of specialized institutions. In: Chebotarev, D. F., ed. *Gerontology and geriatrics. 1982 yearbook. Elderly man. Medical and social care.* Kiev, Gerontological Institute of the USSR Academy of Medical Sciences, 1982, pp. 153–156 (in Russian)

3 BRENNER, M. H. Impact of social and industrial changes on psychopathology: a view of stress from the standpoint of macro societal trends. In: Levi, L., ed. *Society, stress and disease: working life.* Oxford, New York, and Toronto, Oxford University Press, 1981, Vol. 4, pp. 249–260.

4 BROD, J. The influence of higher nervous processes induced by psychosocial environment on the development of essential hypertension. In: Levi, L., ed. *Society, stress and disease: the psychosocial environment and psychosomatic diseases.* London, New York, and Toronto, Oxford University Press, 1971, Vol. 1, pp. 312–323.

5 CEDERBLAD, M. & HÖÖK, B. Day care for three-year-olds. An interdisciplinary experimental study. In: Anthony, E. J. & Chiland, C., ed. *The child in his family: children in turmoil, tomorrow's parents.* New York, Wiley, 1982, Vol. 7, pp. 129–144.

6 COBB, S. & ROSE, R. M. Hypertension, peptic ulcer and diabetes in air traffic controllers. *Journal of the American Medical Association,* **224**: 489–492 (1973).

7 COOPER, C. L. & PAYNE, R., ed. *Current concerns in occupational stress.* Chichester, New York, Brisbane, and Toronto, Wiley, 1980.

8 CULLEN, J. H. & RYAN, G. M. *Occupational factors and health: a review of current issues.* Dublin, Irish Foundation for Human Development, 1981.

9 DENOLIN, H., ed. *Psychological problems before and after myocardial infarction.* Basel, Karger, 1982 (Advances in cardiology, Vol. 29).

10 ELLIOTT, G. R. & EISDORFER, C., ed. *Stress and human health: analysis of implications of research:* New York, Springer, 1982.

11 FLETCHER, B. C. & PAYNE, R. L. Stress at work: a review and theoretical framework, Part 1. *Personnel review,* **9**: 19–29 (1980).

12 GARDELL, B. & JOHANSSON, G., ed. *Working life.* Chichester, New York, Brisbane, and Toronto, Wiley, 1981.

13 GRÖNQVIST, J. Hypertonins sociala orsaker. [Social causes of hypertension.] *Läkartidningen*, **74**: 3968–3970 (1977) (in Swedish with an abstract in English).

14 HAMR, V. Hypertense u zeleznicaru. [Hypertension in railway personnel.] *Pracovní lékařství*, **8**: 126–128 (1956) (in Czech with abstracts in English and Russian).

15 HENRY, J. P. & STEPHENS, P. M. *Stress, health and the social environment: a sociobiologic approach to medicine.* New York Springer, 1977.

16 HOUSE, J. S. Occupational stress and coronary heart disease: a review and theoretical integration. *Journal of health and social behavior*, **15**: 12–27 (1974).

17 KAHN, R. L. *Work and health.* Chichester, New York, Brisbane, and Toronto, Wiley, 1981.

18 KARASEK, R. ET AL. Job, psychological factors and coronary heart disease. Swedish prospective findings and United States prevalence findings using a new occupational inference method. In: Denolin, H., ed. *Psychological problems before and after myocardial infarction.* Basel, Karger, 1982, pp. 62–67 (Advances in cardiology, Vol. 29).

19 KASL, S. V. Epidemiological contributions to the study of work stress. In: Cooper, C. L. & Payne, R., ed. *Stress at work.* Chichester, New York, Brisbane, and Toronto, Wiley, 1978, pp. 3–48.

20 LAPIN, B. A. & CHERKOVICH, G. M. Environmental changes causing the development of neuroses and corticovisceral pathology in monkeys. In: Levi, L., ed. *Society, stress and disease: the psychosocial environment and psychosomatic disease.* London, New York, and Toronto, Oxford University Press, 1971, Vol. 1, pp. 266–279.

21 LAWRENCE, W. G. ET AL. *Physical and psychological stress at work.* Dublin, The European Foundation for the Improvement of Living and Working Conditions, 1982.

22 LEVI, L. Sympatho-adrenomedullary and related biochemical reactions during experimentally induced emotional stress. In: Michael, R. P., ed. *Endocrinology and human behaviour.* London, Oxford University Press, 1968, pp. 200–219.

23 LEVI, L., ed. *Society, stress and disease: the psychosocial environment and psychosomatic diseases.* London, New York, and Toronto, Oxford University Press, 1971, Vol. 1.

24 LEVI, L. Stress and distress in response to psychosocial stimuli. *Acta medica Scandinavica*, Suppl. 528, 1972, Vol. 191.

25 LEVI, L. Psychosocial factors in preventive medicine. In: *Healthy people. The Surgeon General's report on health promotion and disease prevention: background papers.* Washington, DC, United States Government Printing Office, 1979, pp. 207–253 (DEHW Publication No. (PHS) 79-55071A).

26 LEVI, L., ed. *Society, stress and disease: working life.* Oxford, New York, and Toronto, Oxford University Press, 1981, Vol. 4.

27 LEVI, L. *Preventing work stress.* Reading, MA, Addison-Wesley, 1981.

28 LEVI, L. ET AL. Work stress related to social structure and processes In: Elliott, G R & Eisdorfer, C., ed. *Stress and human health: analysis of implications of research.* New York, Springer, 1982, pp. 119–146.

29  MIASNIKOV, A. L. Epidemiology of essential hypertension: discussion. The significance of disturbances of higher nervous activity in the pathogenesis of hypertensive disease. In: *Proceedings of the Joint WHO-Czechoslovak Cardiological Society Symposium on the Pathogenesis of Essential Hypertension, Prague, 22–29 May 1960.* Prague, State Medical Publishing House, 1961, pp. 113 and 153–162.

30  NERELL, G. & WAHLUND, I. Stressors and strain in white collar workers. In: Levi, L., ed. *Society, stress and disease: working life.* Oxford, New York, and Toronto, Oxford University Press, 1981, Vol. 4, pp. 120–127.

31  OBRIST, P. A. *Cardiovascular psychophysiology—a perspective.* New York, Plenum Press, 1981.

32  PETERSEN, E., ed. Livskvalitet—baggrund, begreber, måling. [Quality of life— background, concepts, measurement.] *Psychological reports (Aarhus),* **5**: 2 (1980) (in Danish).

33  RAAB, W. Cardiotoxic biochemical effects of emotional-environment stressors— fundamentals of psychocardiology. In: Levi, L., ed. *Society, stress and disease: the psychosocial environment and psychosomatic diseases.* London, New York, and Toronto, Oxford University Press, 1971, Vol. 1, pp. 331–337.

34  RAHE, R. H. Subjects' recent life changes and their near-future illness susceptibility. In: Lipowski, Z. J. & Hanover, N. H., ed. *Psychosocial aspects of physical illness.* Basel, Karger, 1972, pp. 2–19 (Advances in psychosomatic medicine, Vol. 8).

35  RAHE, R. H. The pathway between subjects' recent life changes and their near-future illness reports: representative results and methodological issues. In: Dohrenwend, B. S. & Dohrenwend, B. P., ed. *Stressful life events: their nature and effects.* New York, London, Sydney, and Toronto, Wiley, 1974, pp. 73–86.

36  RAHE, R. H. & ARTHUR, R. J. Life change and illness studies. *Journal of human stress,* **4**: 3–15 (1978).

37  ROSENMAN, R. H. ed. *Psychosomatic risk factors and coronary heart disease: indications for specific preventive therapy.* Bern, Huber, 1983.

38  ROZWADOWSKA-DOWZENKO, M. ET AL. Nadciśnienie tętnicze samoistne a wykonywany zawod. [Essential hypertension and profession.] *Polskie archiwum medycyny wewnętrznej,* **26**: 497 (1956) (in Polish).

39  SELYE, H. The evolution of the stress concept–stress and cardiovascular disease. In: Levi, L., ed. *Society, stress and disease: the psychosocial environment and psychosomatic diseases.* London, New York, and Toronto, Oxford University Press, 1971, Vol. 1, pp. 299–311.

40  SHARIT, J. & SALVENDY, G. Occupational stress: review and reappraisal. *Human factors,* **24**: 129–162 (1982).

41  Sweden. National Board of Health and Welfare. *Psykisk hälsovård—forskning, social rapportering, dokumentation och information.* [Mental health protection and promotion—research, monitoring, documentation and information.] Stockholm, Liber förlag, 1978 (in Swedish).

42  THEORELL, T. ET AL. A longitudinal study of 21 subjects with coronary heart disease: life changes, catecholamine excretion and related biochemical reactions. *Psychosomatic medicine,* **34**: 505–516 (1972).

43  UNITED STATES OF AMERICA. President's Commission on Mental Health. *Report to the President.* Washington, DC, United States Government Printing Office, 1978, Vol. 1.

44  WAHLUND, I. & NERELL, G. *Work environment of white collar workers. Work, health, wellbeing.* Stockholm, Central Organization of Salaried Employees in Sweden (TCO), 1976.

45  WOLF, S. Psychosocial forces in myocardial infarction and sudden death. In: Levi, L., ed. *Society, stress and disease: the psychosocial environment and psychosomatic diseases.* London, New York, and Toronto, Oxford University Press, 1971, Vol, 1, pp. 324–330.

46  WOLF, S. *Social environment and health.* Seattle and London, University of Washington Press, 1981.

47  WOLF, S. ET AL. *Occupational health as human ecology.* Springfield, IL, Thomas, 1978.

# Chapter 9

# Stress reactions in white- and blue-collar workers

Cary L. Cooper[1]

Until recently the commonly held belief in regard to stress-related illness, in terms of occupation and work, was that it is found predominantly among white-collar "professionals"—i.e., that it is a "bosses' disease". The purpose of this chapter is to lay that myth firmly to rest and to put forward the view that work pressure affects all workers and that the sources and manifestations of stress vary from occupation to occupation.

Table 2 shows that the major causes of death in the working population, such as ischaemic heart disease, often shown to be stress-related, and other illnesses—e.g., pneumonia, prostate cancer—increase across the spectrum of professional–white-collar–unskilled jobs. These statistics from the United Kingdom are very similar to mortality data from the United States of America and other developed countries. In regard to almost all the major, and many of the minor, causes of death, workers in the blue-collar and unskilled groups, are shown to be at greater risk than those in the white-collar and professional groups. This is illustrated not only by mortality statistics but also by morbidity data.

Table 3 shows that in many blue-collar occupations there are greater numbers of restricted-activity days and consultations with general practitioners than in white-collar occupations. It is known that workers' different occupations expose them to health hazards of a physical and chemical nature to varying degrees. Selection into and away from an occupation also plays an important role in the health status of a particular occupational group. Whether professional and managerial workers are less stress prone, their types of occupation and life-style minimizing their vulnerability to stressors at work (and at home) and, consequently, to minor and serious illness, is a question to be considered.

In a large-scale study on stress, anxiety, and work among 1415 men, it was found that a higher proportion of professional and white-collar workers reported nervous strain at work than skilled, semi-skilled, or unskilled manual workers (2). In terms of United Kingdom Office of Population Censuses and Surveys occupational categories the figures were: professional, 53.8%; intermediate non-manual, 56.9%; skilled non-manual, 44.3%; semi-skilled non-manual, 50%; skilled manual, 30.5%; semi-skilled manual, 15.3%; and unskilled manual, 10.3%. Although there may be differences among occupational groups in ways of

[1] Department of Management Sciences, University of Manchester, Manchester, England.

Table 2. Deaths by major causes and types of occupation, 1970–1972[a]
(Standardized mortality rate=100)

| Causes of death (males aged 15–64 years) | Professional and similar | Intermediate | Skilled non-manual | Skilled manual | Partly skilled | Unskilled |
|---|---|---|---|---|---|---|
| Trachea, bronchus, and lung cancer | 53 | 68 | 84 | 118 | 123 | 143 |
| Prostate cancer | 91 | 89 | 99 | 115 | 106 | 115 |
| Ischaemic heart disease | 88 | 91 | 114 | 107 | 108 | 111 |
| Other forms of heart disease | 69 | 75 | 94 | 100 | 121 | 157 |
| Cerebrovascular disease | 80 | 86 | 98 | 106 | 111 | 136 |
| Pneumonia | 41 | 53 | 78 | 92 | 115 | 195 |
| Bronchitis, emphysema, and asthma | 36 | 51 | 82 | 113 | 128 | 188 |
| Accidents, other than motor vehicle | 58 | 64 | 53 | 97 | 128 | 225 |
| All causes | 77 | 81 | 99 | 106 | 114 | 137 |

[a] Source: United Kingdom of Great Britain and Northern Ireland, Office of Population Censuses and Surveys.

Table 3. Days of acute sickness and number of consultations with general medical practitioners, 1974–1975[a]

| Type of occupation | Average number of restricted-activity days per person (males) per year | | | Average number of consultations per person (males) per year | | |
|---|---|---|---|---|---|---|
| | Age group (years) | | All ages | Age group (years) | | All ages |
| | 15–44 | 45–64 | | 15–44 | 45–64 | |
| Professional | 9 | 16 | 12 | 2.1 | 2.7 | 2.7 |
| Employers and managers | 11 | 13 | 14 | 1.8 | 2.4 | 2.7 |
| Intermediate and junior non-manual | 10 | 21 | 15 | 2.0 | 4.3 | 3.1 |
| Skilled manual and own account non-professional | 15 | 24 | 17 | 2.8 | 4.0 | 3.2 |
| Semi-skilled manual and personal service | 16 | 23 | 18 | 2.7 | 4.5 | 3.7 |
| Unskilled manual | 21 | 28 | 20 | 3.5 | 4.8 | 3.6 |
| All persons | 13 | 21 | 16 | 2.4 | 3.8 | 3.1 |

[a] Source: United Kingdom of Great Britain and Northern Ireland, Office of Population Censuses and Surveys. (General household survey, 1974 and 1975).

reporting perceived symptoms, this may also indicate that white-collar and professional workers differ from blue-collar workers in their reactions to stress—i.e., in the former the pressures of work may be reflected to a larger extent by psychological problems and mental illness, and in the latter by physical symptoms and illness. There may, however, be differences from occupation to occupation within both the white- and the blue-collar categories. This was shown in a large-scale study carried out by a team of research workers at the Institute of Social Research, University of Michigan in which over 2000 men employed in 23 occupational groupings—from fork-lift driver, paced-assembly line operator, and electrical technician, to policeman, air traffic controller, accountant, professor, and physician—were examined (1). A variety of differences within blue-collar occupations and within white-collar occupations were found in the crude measures of stress manifestations and illness. For example, in the blue-collar occupations, although a high incidence of self-reported illness and somatic complaints was found among machine-paced assembly line workers, it was low among continuous-flow workers. In the white-collar occupations, a high incidence of somatic complaints was found among air traffic controllers while it was low among professional engineers and computer programmers. Nevertheless, the indices of ill health were greater for the blue-collar workers than for the white-collar workers.

The value of the University of Michigan study was that it enabled the sources of stress or the stressors among the 23 occupational groupings to be determined. Three major stressors differentiated white-collar and professional workers, such as administrators, doctors, teachers, and air traffic controllers, from blue-collar workers: high and variable work-load, responsibility for people, and job complexity and demands for concentration. Different stressors were related to ill health among blue-collar workers: uncertainty about future employment, the under-utilization of abilities, role ambiguity, and lack of job complexity. These findings draw attention to two important aspects of occupational wellbeing.

First, if a variety of white-collar jobs are examined in detail, job dissatisfaction and stress seem to result primarily from too little delegation and decentralization of tasks and decision-making. On the other hand, many blue-collar workers—e.g., machine-paced assembly line workers, tenders of machines— seem to suffer from quite the opposite, experiencing as stressors lack of job complexity and under-utilization of skills and abilities. This may be the reason why there is a growing movement in Western Europe and North America towards greater industrial democracy at work—e.g., autonomous work groups and group decision-making—and the re-designing of jobs. It could be that, in a wide variety of occupations, the delegation of decision-making would not only alleviate stress among many white-collar and professional workers but also encourage a greater utilization of abilities and enhance job satisfaction among blue-collar workers. It may, therefore, be possible to understand from the University of Michigan study the current world-wide trends in the humanization of work, particularly as they concern the re-designing of jobs and greater participation in decision-making from the shop floor.

Second, results from a number of studies indicate that, to reduce stress at work and to maximize job satisfaction, an extensive "stress audit" of particular jobs has to be carried out, built on a medical model, with a careful and detailed diagnosis of each aspect of a particular job, isolating its stress debits and its stress credits (3, 4). On the basis of such an audit, plans for change, both organizational and in relation to the job itself, can be made, with the involvement of the workers. To begin to use stress prevention techniques, such as transcendental meditation, transactional analysis, or stress counselling, before a thorough and systematic audit of a particular job has been carried out is unwarranted. All too often remedies or currently in vogue training approaches are applied to problems that have not had their causes adequately diagnosed. It is becoming necessary to expend fresh effort in relation to stress at work since, as has been suggested by Cox, "The goal of all concerned with industry, it is now widely agreed, is the improvement in the quality of working life. Occupational stress is the threat to work" (5).

## References

1 CAPLAN, R. D. ET AL. *Job demands and worker health: main effects and occupational differences.* Washington, DC, United States Government Printing Office, 1975 (DEHW Publication No. (NIOSH) 75-160).

2 CHERRY, N. Stress, anxiety and work. *Journal of occupational psychology,* **51**: 259–270 (1978).

3 COOPER, C. L. *Stress research: issues for the eighties.* London and New York, Wiley, 1982.

4 COOPER, C. L. & PAYNE, R. *Stress at work.* Chichester, New York, Brisbane, and Toronto, Wiley, 1978.

5 Cox, T. *Stress.* London, Macmillan, 1978.

# Chapter 10

# Sources of stress at work and their relation to stressors in non-working environments

Cary L. Cooper and Marilyn Davidson[1]

## Introduction

Sources of occupational stress can only be adequately investigated by using a multidisciplinary approach—i.e., by examining the whole spectrum of psychological, sociological, and physiological problems that make stimulus demands on an individual in his working environment.

Use of a multidisciplinary approach acknowledges also that stressors in the working environment can affect an individual at home and in his social environment and vice versa. Thus, when studying the sources and manifestations of stress in a specific occupational group—e.g., personnel in factories or hospitals—it is essential to be aware of the importance of the extra-organizational sources of stress that can affect the performance and the mental and physical health of an individual at work. Two major extra-organizational areas from within which stress may arise are the home environment—e.g., marital relations and financial concerns—and the social environment—e.g., social activities and relationships and urban versus rural living conditions.

A large number of potential stressors in the working environment have been determined in numerous studies. They include such factors as shift work; the under-utilization of abilities; work underload and boredom; work overload; role conflict; unequal pay; job future ambiguity; relationships at work; the quality of equipment; and physical danger. These factors may contribute to a number of detrimental outcomes, including behavioural outcomes, such as impaired job performance, alcohol abuse, cigarette smoking, and drug abuse; to physical illness, such as migraine and hypertension; and to mental illness, such as depression.

The areas of life that constitute possible sources of stress and subsequent stress outcomes are the work arena, the home arena, the social arena, and the individual arena. A comprehensive multifaceted model of occupational stressors and stress outcomes, making up an interrelated whole, is presented in Fig. 3. The three arenas for behaviour—i.e., the work, home, and social environments—are composed of forces that impinge upon, and interact with, the individual. Thus, a stressor from one arena

[1] Department of Management Sciences, University of Manchester, Manchester, England.

## Fig. 3. A model of occupational stress

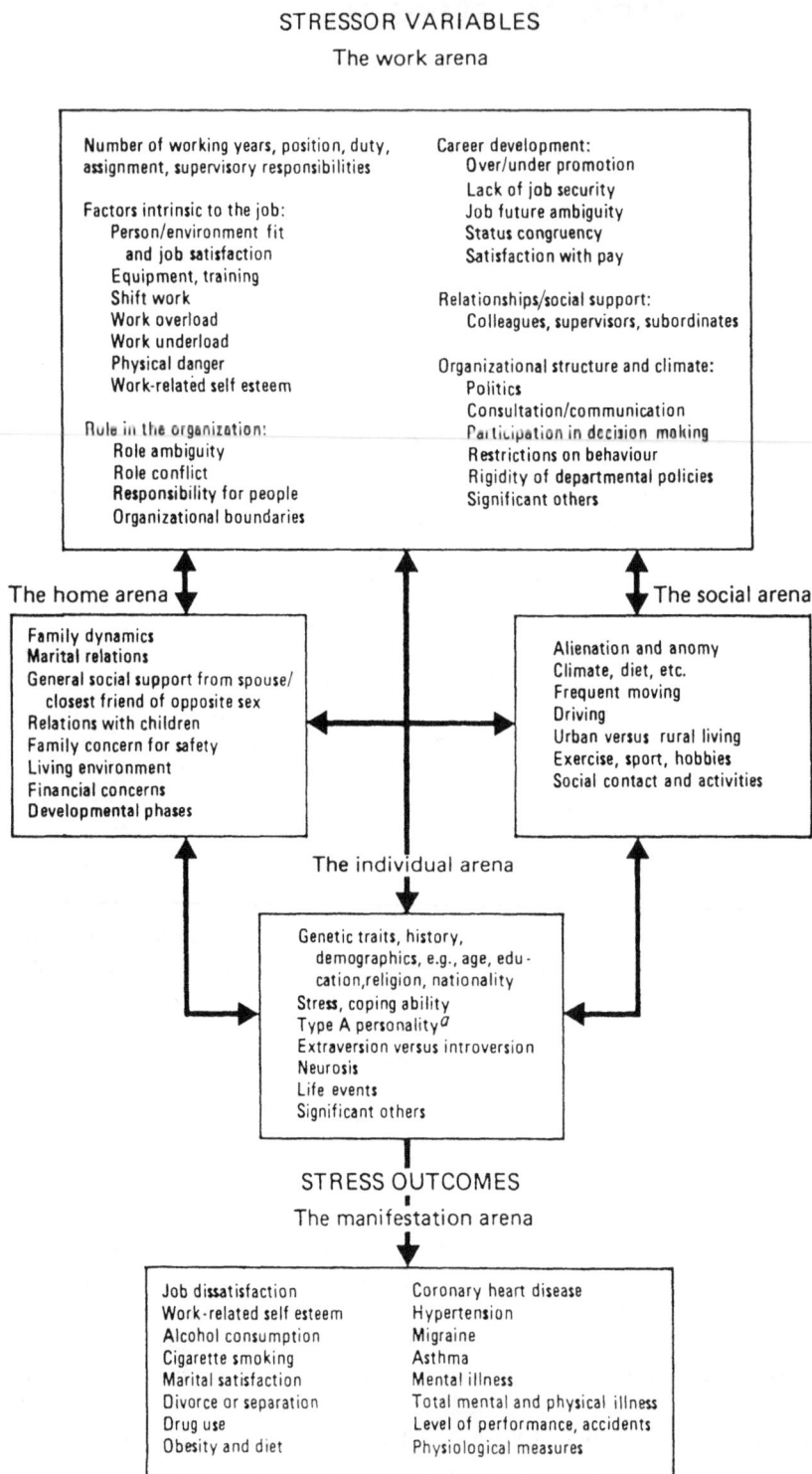

### STRESSOR VARIABLES
#### The work arena

| | |
|---|---|
| Number of working years, position, duty, assignment, supervisory responsibilities | Career development:<br>    Over/under promotion<br>    Lack of job security<br>    Job future ambiguity<br>    Status congruency<br>    Satisfaction with pay |
| Factors intrinsic to the job:<br>    Person/environment fit<br>      and job satisfaction<br>    Equipment, training<br>    Shift work<br>    Work overload<br>    Work underload<br>    Physical danger<br>    Work-related self esteem | Relationships/social support:<br>    Colleagues, supervisors, subordinates |
| Role in the organization:<br>    Role ambiguity<br>    Role conflict<br>    Responsibility for people<br>    Organizational boundaries | Organizational structure and climate:<br>    Politics<br>    Consultation/communication<br>    Participation in decision making<br>    Restrictions on behaviour<br>    Rigidity of departmental policies<br>    Significant others |

**The home arena**

Family dynamics
Marital relations
General social support from spouse/
    closest friend of opposite sex
Relations with children
Family concern for safety
Living environment
Financial concerns
Developmental phases

**The social arena**

Alienation and anomy
Climate, diet, etc.
Frequent moving
Driving
Urban versus rural living
Exercise, sport, hobbies
Social contact and activities

**The individual arena**

Genetic traits, history,
    demographics, e.g., age, edu-
    cation, religion, nationality
Stress, coping ability
Type A personality[a]
Extraversion versus introversion
Neurosis
Life events
Significant others

### STRESS OUTCOMES
#### The manifestation arena

| | |
|---|---|
| Job dissatisfaction | Coronary heart disease |
| Work-related self esteem | Hypertension |
| Alcohol consumption | Migraine |
| Cigarette smoking | Asthma |
| Marital satisfaction | Mental illness |
| Divorce or separation | Total mental and physical illness |
| Drug use | Level of performance, accidents |
| Obesity and diet | Physiological measures |

WHO 851815

[a] See Chapter 13.

can affect the individual and this, in turn, may activate changes in the type and force of a stressor from another arena.

When an individual experiences stress to a sufficient extent in one (or more) arena it will become manifest. The specific nature of the manifestation depends on a number of variables within the context of the individual situation. Thus, when exposed to apparently the same stressor, one individual may engage in escapist drinking while another may develop bronchitis.

The following review will relate to each of the three arenas for behaviour and their subsystems within the model, enabling multidisciplinary approaches to occupational stress to be advocated.

## The social and cultural arena

In research studies that concentrate on social and cultural factors related to stress, incidence of chronic disease and mortality rates tend to be used as the main measurement variables (14, 27). In a study published in 1974 the extent of socially induced stress in a population was viewed as being dependent on such variables as status integration, role conflict, frustration, and goal attainment (18). In another report the influence of advertising, the mass media, and other commercial forces on affluent industrial societies was discussed and the hypothesis presented that their interaction builds up demands and expectations (overstimulation) in different areas of life, which can affect health (19). It was maintained that this can be substantiated by the fact that affluent industrial societies have higher incidences of chronic stress-related disease, such as coronary heart disease, than do less affluent societies.

### Alienation and anomy

It has been proposed that the complex need-structure and the societal pressures inherent in affluent societies have particularly adverse effects on workers in certain occupations—e.g., the police—resulting in such states as role conflict, powerlessness, alienation, and anomy (13, 26). Risk of anomy and alienation is especially great among migrant workers, who are coming to form a considerable proportion of the labour force throughout the world. Feelings of alienation can lead to the state of anomy, a term coined by Durkheim to describe conditions of normlessness. Adverse feelings of alienation and anomy can develop in specific occupational groups who experience isolation from the community and/or adverse community relationships (*12, 39*). Furthermore, the concept of anomy has been postulated as one of the major influencing variables in the psychosocial theory of suicide (*15*). This has recently been verified by Wenz, who found that anomy was positively related to degree of lethality, being highest in serious suicide-attempt subjects (*43*).

## Climate, diet, etc.

It has been postulated that social and cultural stressors are influenced by factors such as diet, climate, genetics, religion, social class, overcrowding, type of dwelling, and isolation, all of which vary with culture and geography (*37*). For instance, studies have shown that arteriosclerotic coronary heart disease is relatively uncommon in Japan compared with the white population of North America (*30*).

## Frequent moving

Another variable affecting life-style and relationships is frequent moving. Families who move frequently are unable to make meaningful ties with local communities (*9*). This can enhance feelings of isolation and cause stress for all the family members. Unfortunately, moving frequently is an occupational hazard for many of those pursuing careers. A study of a rural population showed that some factors associated with people who tend to move frequently from one social setting or occupation to another are especially conducive to coronary heart disease (*41*).

## Driving

Means of transport to and from work appears to be a relevant variable, especially for high risk individuals (*4*). For example, excretion of both 11-hydroxycorticosteroid and catecholamines was found to increase significantly during a 2-hour period of driving in both healthy subjects and subjects suffering from coronary artery disease (*2*).

## Urban versus rural living conditions

Demographic variation relative to the prevalence of stress disorders has been found to exist between urban and rural communities. It was found that the prevalence of hypertension is often higher in urban than in rural communities (27) and that coronary heart disease is persistently higher along the eastern and western seaboards of North America than in the inland rural farm areas (14). California, for instance, has nearly twice the death rate from arteriosclerotic heart disease of nearby rural Mexico.

It has, however, been proposed that, although certain conditions, such as hypertension, are often more prevalent in urban than in rural areas, the factors associated with city living—e.g., noise, overcrowding, and pace—could just as easily be counteracted by stressful factors in country living, e.g., isolation and lack of privacy and social control (27).

The long-held belief that better mental health is experienced by rural dwellers in comparison with urban dwellers was challenged after a 20-year follow-up of an original study carried out in Manhattan, USA. The results of that study were compared with the results of a nation-wide investigation in the USA from 1960 to 1962, of the mental health of 6700 adults living in communities ranging from 3 million to less than 2500 (40). The findings were based on self-reported symptoms—e.g., having had a nervous breakdown—and showed that those living in communities of less than 50 000 population had a 20% higher symptom score than those living in communities of over 50 000. Similar outcomes were reported in New Zealand, where it was found that the number of psychotropic drug prescriptions issued in the rural areas was double the number issued in the urban areas (40). It is apparent that the literature on health differences between urban and rural populations is inconsistent; it is still difficult to define causal relationships demonstrating why the differences occur.

## The home arena

When investigating the disruption of social, home, and family life as a direct outcome of stress resulting from occupational stressors a multitude of interrelating variables must be taken into consideration. Firstly, there is a feedback loop, with stressors at work being able to affect family life, and vice versa, one often exacerbating the other (8, 9). Such a feedback loop was, however, denied by Croog who suggested that the family itself is never a source of stress but a "crystallizing entity" within which external stressors "emerge and exert their impact" (27).

When attempting to investigate the correlation between occupational stress and its effect on family life, it is important to be aware that Rollins and Feldman's index of marital satisfaction shows the trend as a U-shaped curve, with the decline in the middle years of married life and the increase in the later years (35). It was suggested that role-strain intervenes between family life cycle and marital satisfaction, thereby making it a critical variable in explaining the U-shaped curve type of relationship between the two (34). Individuals' roles within the family result in more conflict during the middle years of the family life cycle, especially as regards the time spent in working if it interferes with the role within the family.

Furthermore, a period of particularly high risk in marital life is when the children are at school. A major problem at that time is the role-conflict between husband and wife, often resulting in family anxiety and the inability of the husband to understand the wife's problems. This is the case especially if he is establishing his career and to him the family life appears stable. When the children are in their teens, infidelity on the part of the husband or feelings of inadequacy and failure if he has not realized his life ambitions, including promotion at work, are two major causes of marriage conflict (25). Therefore, it can be concluded that most marital conflict occurs in the middle years of the marital life cycle. Obviously if an individual is also experiencing work stress during this high marital risk period the conflict will increase.

A factor related to the home and social environment that has been associated with stress outcome is the absence of social support (3, 28). Lack of job-related social support—i.e., support from supervisors and co-workers—has been shown to be a potential occupational stressor (3), and some studies have illustrated that the absence of supportive friends and family is important as a predictor of emotional disturbance (21, 36).

As well as leading to emotional disturbance, inadequate social support, especially what could be expected from a wife, has been related to stress-induced physical illness. In a study of 1809 white, male, blue-collar workers, social support from wives and supervisors was found to be more important than support from relatives, friends, or co-workers, in protecting them from developing ulcers in response to perceived stress (28).

## The work arena

Having discussed a range of extra-organizational stressors that it is essential to take into consideration in occupational stress research, the literature delineating specific organizational stressors will be briefly reviewed. The following five major sources of occupational stress will be discussed: (*a*) factors intrinsic to the job; (*b*) role in the organization; (*c*) career development; (*d*) relationships at work; and (*e*) organizational structure and climate.

## Factors intrinsic to the job

In a variety of occupations, sources of stress intrinsic to the job include poor physical working conditions, shift work, work overload, work underload, and physical danger (*6, 7, 8*).

### Ergonomic conditions

Poor physical working conditions can exacerbate stress at work. For example, it is thought that the design of the control room at a nuclear power plant is an important variable in terms of stress to the workers and that more sophisticated ergonomic designs are required (*31*). It was found that an important stress factor leading to the Three Mile Island nuclear power plant accident was the distraction caused by the excessive sounding of emergency alarms (*31*). Air traffic controllers frequently complain about ergonomic conditions at work (*11*). Most ergonomically deprived environments, however, are those of blue-collar workers (*33*).

### Shift work

In numerous occupational studies it has been found that shift work is a common occupational stressor, affecting neurophysiological rhythms, such as body temperature, metabolic rate, and blood sugar level, mental efficiency, and work motivation, which may ultimately result in stress-related disease (*5, 37*). Air traffic control is a particularly highly stressed occupation; four times the prevalence of hypertension and also more mild diabetes and peptic ulcer was found in a group studied in the early 1970s than in its control group of second class airmen. Although other job stressors were attributed as being instrumental in the causation of these stress-related maladies, shift work was the major one isolated (*5*).

Nevertheless, although it is acknowledged that there are stressors associated with shift work, Selye suggests that most investigations support the conclusion that shift work becomes physically less stressful as individuals adapt to it (*37*). Even so, "exclusion from society" is a common complaint among shift workers.

## Work overload

Work overload is seen as being either quantitative—i.e., too much to do—or qualitative—i.e., too difficult (20). Although empirical evidence demonstrating that work overload is a main factor in occupational ill health is not available, it has been associated with certain behavioural malfunctions (8, 9, 23). For example, a relationship was demonstrated between quantitative overload and cigarette smoking, an important risk factor in coronary heart disease (20); and in a study of 1500 employees job overload was found to be associated with such stress-related symptoms as lowered self-esteem, low work motivation, and escapist drinking (29).

## Work underload

Work underload together with a repetitive, routine, boring, or understimulating working environment—e.g., a paced assembly line—has been associated with ill health (10). This may, for example, be a problem facing a nuclear power plant operator, as 99.9% of his time is spent on monotonous, rather than stimulating, tasks (31). Indeed, as in the job of policeman, an operator in a nuclear power plant has to accept periods of boredom that may suddenly be disrupted because of an emergency; this sudden jolt to the physical and mental state may have a detrimental effect on his health (16). Furthermore, boredom and lack of interest in the job may reduce an operator's ability to respond to an abnormal situation (31).

## Physical danger

Certain occupations have been determined to be of high risk in terms of danger—e.g., policeman, miner, soldier, and fireman (13, 22). When faced with abnormal situations, nuclear power plant personnel are also subjected to physical danger but the type of stress that is induced by uncertainty as to whether an event presenting physical danger might occur, is often substantially relieved if the worker feels himself to be adequately trained and equipped to cope with the emergency should it arise (12).

# Role in the organization

It has been determined that a person's role at work is a main source of occupational stress. Stress may stem from role ambiguity—i.e., not being clear as to what is required—or role conflict—i.e., conflicting job demands—such as are involved in being responsible for people, and conflict arising from organizational boundaries (8). It has been indicated that organizational stressors stemming from role ambiguity and conflict can result in stress-related maladies such as coronary heart disease (1, 20, 38). Furthermore, it has been found that those in managerial, clerical, and professional occupations are more prone to the type of occupational stress that is related to role conflict (8).

After a review of the relevant literature it was concluded that the correlations between role conflict and ambiguity and the components of job satisfaction tend to be strong; between role conflict and ambiguity and mental disorder, however, they tend to be weak (23). Personality is an important determinant of how an individual reacts to role conflict; greater job-related tension is produced in introverts than in extroverts and it is held that flexible people show greater job-related tension under conditions of conflict than do rigid individuals (20).

Responsibility for workers and for their safety appears to be a potentially viable occupational stressor. The pressures of nuclear power plant operators were defined in terms of their responsibility for the safety of others when faced with abnormal situations (31). Responsibility for others was seen as a potential stressor in police work although not to the same extent as in air traffic control (24). This was verified by a study of occupational stress in air traffic controllers which determined responsibility for people's safety and lives to be a major occupational stressor (11).

## Career development

Cooper & Marshall maintain that environmental stressors related to career development stem from "the impact of overpromotion, underpromotion, status incongruence, lack of job security, thwarted ambition, etc" (8). In a study of United States Navy employees, status congruency, or the degree to which there is job advancement, including promotion to the next rank, was found to be positively related to military effectiveness and negatively related to the incidence of psychiatric disorder (17). However, it was considered, in the case of nuclear power plant operators, that large increases in pay would not necessarily increase their job satisfaction; the result could be that they would remain in jobs that no longer gave them satisfaction (31).

Relationships at work

Relationships at work, their nature, and the social support received from colleagues, supervisor, and subordinates, have been related to job stress (32). Poor relations with other members of an organization may be precipitated by role ambiguity, which produces psychological strain in the form of low job satisfaction (20). It was found that a high degree of social support from peers relieves job strain; it was also shown to condition the effect of job stress on cortisone, blood pressure, and glucose levels; and the number of cigarettes smoked as well as the rate of giving up cigarette smoking (3). It is interesting to note that greater help and social support is provided to air traffic controllers by friends and colleagues than by supervisors (11).

Organizational structure and atmosphere

Occupational stress in relation to organizational structure and atmosphere results from such factors as office politics, lack of effective consultation, exclusion from the decision-making process, and restrictions on behaviour (9, 42). It was found that greater participation led to higher productivity, improved performance, lower staff turnover, and lower levels of physical and mental disorder, including such stress-related behaviour as escapist drinking and heavy smoking (20, 29).

## Conclusions

In this review and analysis of the literature related to the sources of occupational stress, the focus has been on stress as a multifaceted, multidimensional construct which can be fully understood only if a thorough examination is made of all the social and physical arenas in which an individual lives. The stress, health, job performance, family, and social network, with individual differences, forms an integrated whole. Fig. 3 illustrates this concept.

# References

1  BEEHR, T. A. ET AL. Relationship of stress to individually and organisationally valued states: higher order needs as a moderator. *Journal of applied psychology,* **61**: 41–47 (1976).

2  BELLET, S. ET AL. The effect of automobile driving on catecholamine and adrenocortical excretion. *American journal of cardiology,* **24**: 365–368 (1969).

3  CAPLAN, R. D. ET AL. *Job demands and worker health: main effects and occupational differences.* Washington, DC, United States Government Printing Office, 1975 (DEHW Publication No. (NIOSH) 75-160).

4  CARRUTHERS, M. Hazardous occupations and the heart. In: Cooper, C. L. & Payne, R., ed. *Current concerns in occupational stress.* Chichester, New York, Brisbane, and Toronto, Wiley, 1980, pp. 3–22.

5  COBB, S. & ROSE, R. M. Hypertension, peptic ulcer and diabetes in air traffic controllers. *Journal of the American Medical Association,* **224**: 489–492 (1973).

6  COOPER, C. L. *The stress check.* Englewood Cliffs, NJ, Prentice-Hall, 1980.

7  COOPER, C. L. *Executive families under stress.* Englewood Cliffs, NJ, Prentice-Hall, 1981.

8  COOPER, C. L. & MARSHALL, J. Occupational sources of stress. A review of the literature relating to coronary heart disease and mental ill health. *Journal of occupational psychology,* **49**: 11–28 (1976).

9  COOPER, C. L. & MARSHALL, J. Sources of managerial and white collar stress. In: Cooper, C. L. & Payne, R., ed. *Stress at work.* Chichester, New York, Brisbane, and Toronto, Wiley, 1978, pp. 81–105.

10  COX, T. Repetitive work. In: Cooper, C. L. & Payne, R., ed. *Current concerns in occupational stress.* Chichester, New York, Brisbane, and Toronto, Wiley, 1980, pp. 23–41.

11  CRUMP, J. H. ET AL. Investigating occupational stress: a methodological approach. *Journal of occupational behaviour,* **1**: 191–202 (1980).

12  DAVIDSON, M. J. *Stress in the police service: a multifaceted model, research proposal and pilot study.* Thesis, University of Queensland, Australia, 1979.

13  DAVIDSON, M. J. & VENO, A. Stress and the policeman. In: Cooper, C. L. & Marshall, J., ed. *White collar and professional stress.* London, Wiley, 1980, pp. 131–166.

14  DODGE, D. L. & MARTIN, W. T. *Social stress and chronic illness— mortality patterns in industrial society.* Notre Dame, IN, University of Notre Dame Press, 1970.

15  DURKHEIM, E. *Suicide.* New York, The Free Press, 1951.

16  EISENBERG, T. Labor–management relations and psychological stress—view from the bottom. *The police chief,* **42**: 54–58 (1975).

17  ERICKSON, J. M. ET AL. Status congruency as a prediction of job satisfaction and life stress. *Journal of applied psychology,* **56**: 523–525 (1972).

18 FRANKENHAEUSER, M. *Man in technological society: stress, adaptation and tolerance limits.* Reports from the Psychological Laboratories, University of Stockholm, Suppl. 26, 1974.

19 FRANKENHAEUSER, M. *Quality of life: criteria for behavioural adjustment.* Reports from the Department of Psychology, University of Stockholm, 1976 (Report No. 475).

20 FRENCH, J. R. P. JR. & CAPLAN, R. D. Organizational stress and individual strain. In: Marrow, A. J., ed. *The failure of success.* New York, AMACOM, 1972, pp. 30–66.

21 GORDON, R. & GORDON, K. Social factors in prevention of post partum emotional problems. *Paper presented at the Annual Meeting of the American Orthopsychiatric Association, Washington, DC, 1967.* New York, American Orthopsychiatric Association,1967.

22 KASL, S. V. Mental health and work environment: an examination of the evidence. *Journal of occupational medicine,* **15**: 506–517 (1973).

23 KASL, S. V. Epidemiological contributions to the study of work stress. In: Cooper, C. L. & Payne, R., ed. *Stress at work.* Chichester, New York, Brisbane, and Toronto, Wiley, 1978, pp. 3–48.

24 KROES, W. H. *Society's victim— the policeman: an analysis of job stress in policing.* Springfield, IL, Thomas, 1976.

25 KRUPINSKI, J. & STROLLER, A. *The family in Australia.* Sydney, Pergamon Press, 1974.

26 LEFKOWITZ, J. Industrial organizational psychology and the police. *American psychologist,* **32**: 346–364 (1977).

27 LEVINE, S. & SCOTCH, N. A. ed. *Social stress.* Chicago, IL, Aldine, 1970.

28 McMICHAEL, A. J. Personality, behavioural, and situational modifiers of work stressors. In: Cooper, C. L. & Payne, R., ed. *Stress at work.* Chichester, New York, Brisbane, and Toronto, Wiley, 1978, pp. 127–147.

29 MARGOLIS, B. L. ET AL. Job stress: an unlisted occupational hazard. *Journal of occupational medicine,* **16**: 659–661 (1974).

30 MATSUMOTO, Y. S. Social stress and coronary heart disease in Japan: a hypothesis. *Milbank Memorial Fund quarterly: health and society,* **48**: 9–36 (1970).

31 OTWAY, H. J. & MISENTA, R. The determinants of operator preparedness for emergency situations in nuclear power plants. *International Workshop on Procedural and Organisational Measures for Accident Management: Nuclear Reactors, Laxenburg, Austria, 28–31 January 1980.* Laxenburg, Austria, International Institute for Applied Systems Analysis, 1980.

32 PAYNE, R. Organizational stress and social support. In: Cooper, C. L. and Payne, R., ed. *Current concerns in occupational stress.* Chichester, New York, Brisbane, and Toronto, Wiley, 1980, pp. 269–298.

33 POULTON, E. C. Blue collar stressors. In: Cooper, C. L. & Payne, R., ed. *Stress at work.* Chichester, New York, Brisbane, and Toronto, Wiley, 1978, pp. 51–79.

34 ROLLINS, B. C. & CANNON, K. L. Marital satisfaction over the family life cycle: a re-evaluation. *Journal of marriage and the family,* **36**: 271–283 (1974).

35 ROLLINS, B. C. & FELDMAN, H. Marital satisfaction over the family life cycle. *Journal of marriage and the family*, **32**: 20–28 (1970).

36 SEGAL, B. ET AL. Social integration, emotional adjustments and illness behavior. *Social forces*, **46**: 237–241 (1967).

37 SELYE, H. *Stress in health and disease*. Boston and London, Butterworths, (1976).

38 SHIROM, A. ET AL. Job stresses and risk factors in coronary heart disease among five occupational categories in kibbutzim. *Social science and medicine*, **7**: 875–892 (1973).

39 SKOLNICK, J. A sketch of the policeman's working personality. In: Niederhoffer, A. & Blumberg, A. S. ed. *The ambivalent force*. San Francisco, CA, Rhinehart Press, 1973, pp. 132–143.

40 SROLE, L. & FISHER, A. K. *Mental health in the metropolis: the Midtown Manhattan study*. New York, New York University Press, 1978.

41 SYME, S. L. ET AL. Cultural mobility and coronary heart disease in an urban area. *American journal of epidemiology*, **82**: 334–346 (1966).

42 VENO, A. & DAVIDSON, M. J. A relational model of stress and adaptation. *Man–environment systems*, **8**: 75–89 (1978).

43 WENZ, V. F. Anomie and level of suicidality in individuals. *Psychological reports*, **36**: 817–818, 1975.

Chapter 11

# Physical and chemical factors that increase vulnerability to stress or act as stressors at work

Kari Lindström and Sirkka Mäntysalo[1]

## Introduction

Some physical and chemical factors in the working environment not only affect physical health adversely but can be detrimental to mental health. As well as psychosocial factors, such as those associated with work content and working arrangements, there are industrial chemicals, such as organic solvents and heavy metals, that are neurotoxic and can adversely affect psychological functions and emotional behaviour; and physical factors, such as noise, vibration, and thermal conditions that can also produce changes in psychological functions and emotional reactions. If social-psychological working conditions are to improve, these chemicals and physical factors must be taken into consideration. Even when they do not directly influence the psychological functions and mental health of the workers they may still lead to lowered job satisfaction.

## Harmful physical and chemical agents

In surveys of various occupational groups, complaints often refer to noise, thermal conditions, vibration, and chemicals as the most harmful perceived stressors. In a survey in Finland of blue-collar workers, noise was considered very, or moderately, harmful by 52% of the respondents, and 47% rated thermal conditions the same (19). Often, however, these agents are only assumed stressors concealing the real load factors. One reason for this may be that their concrete nature, and the relatively extensive information available about their adverse effects, can create awareness leading to insecurity among the workers.

An extreme phenomenon related to exposure to physical factors and chemicals is mass psychogenic illness (4). This is characterized by the sudden and dramatic occurrence of subjective nonspecific symptoms attributed to such factors in the working environment as a strange odour or contaminated air. A more detailed study, however, reveals that factors associated with work content or working arrangement are responsible.

[1] Institute of Occupational Health, Helsinki, Finland.

## Organic solvents

In industry, organic solvents are the most common neurotoxic agents. Because there are thousands of such solvents, and because new ones are being developed constantly, knowledge of their adverse effects cannot be complete. They do not constitute a homogeneous group. Many belong to such main groups as the halogenated hydrocarbons (e.g., trichloroethylene and perchloroethylene) the aromatic hydrocarbons (e.g., toluene, styrene, and xylene) and the aliphatic hydrocarbons (such as mineral oils).

In most cases occupational exposure to organic solvents involves a mixture rather than a single solvent. Painting is one of the major sources of exposure. Other examples are dry-cleaning, the degreasing of metals, lamination, gluing, and photogravure printing.

The adverse psychological effects of working with neurotoxic chemicals can be manifested in such functions as the memory and learning, in the sensory and motor functions, and sometimes also in the intellectual functions. Personality characteristics may also be altered.

The first signs of adverse effects are subjective symptoms during the working day, or chronic subjective symptoms. When the uptake of organic solvents is excessive, such acute symptoms as irritation of the mucous membranes and the upper respiratory tract and some neural symptoms usually occur. The most common chronic symptoms are neurasthenic, and include unusual fatigue, headache, sleep disturbance, irritability, and anxiety. Sweating and nausea, associated with an imbalance in the autonomic nervous system, are also common.

Sometimes the use of alcohol is associated with occupational exposure to solvents. There is clinical evidence that the amount of alcohol consumed increases with the duration of exposure to solvents. In some cases, however, the alcohol tolerance of workers exposed to solvents has lowered, and the amount of alcohol consumed has decreased (23). The anecdotal data—e.g., relating to the excessive use of alcohol among painters—may be explained by the addictive effects of long-term solvent exposure. However, in one study house painters' self-reported use of alcohol was the same as the use of alcohol among workers with other occupations in the construction industry (25). The chronic effects on the psychological state of exposure to organic solvents are described below.

## Halogenated hydrocarbons

The psychological effects of long-term exposure to trichloro-ethylene are well documented. The psychoorganic deterioration found was characterized by impairment of cognitive and psychomotor functions and by affective alterations (8, 18). There was a report that long-term exposure to perchloroethylene produced encephalopathy and pseudoneurasthenic symptoms but a controlled field study did not give similar findings (38). Similarly, no definite conclusions can be reached in regard to the neurotoxicity of 1,1,1-trichloroethane after chronic exposure to it, because only one study on the subject has been carried out so far (29).

## Aromatic hydrocarbons

Among the aromatic hydrocarbons, the psychological effects of styrene and toluene have been studied. In studies of workers involved in the lamination process, prolonged reaction times and lowered visuomotor accuracy were observed after long-term exposure to styrene (7, 22). Visuomotor inaccuracy was related to exposure doses. Workers exposed to styrene also showed lowered emotionality in personality measurements (24).

After excessive exposure to toluene (2.6 g/m$^3$) photogravure printing workers displayed psychoorganic deterioration (31). Lower levels of exposure (225–750 mg/m$^3$) have affected narrower areas of psychological functioning, particularly short-term memory (12). In a study of workers exposed to toluene in the Democratic Republic of Germany, the functions most affected were memory and attention (35).

## Paint solvents

In the past, paints composed mainly of mixtures of aromatic hydrocarbons were in common use. Recently, however, they are used less frequently than water-based paints and those containing aliphatic hydrocarbons. Car painters were investigated in a study in Finland (13) and car and industrial painters in a study in Sweden (5). Both cognitive and sensory and motor functions showed deterioration. The study in Finland showed mainly impairments in perceptual organization and verbal memory, whereas in that in Sweden more disturbances were found in the sensory and motor functions. The car painters in Finland also showed lowered emotionality and less control of behaviour and thinking. The lowered emotionality was explained as being associated with exposure, but the behavioural reaction was considered to be more of a secondary reaction to changes in the whole life situation. The estimated levels of chronic exposure were clearly below the applied hygienic limits—e.g., in the case of the car painters in Finland the level was only one-third of the applied value (13).

Two psychological investigations of house painters provided similar results; reaction times were prolonged and visual memory was lowered (10, 25).

## Carbon disulfide

The psychological effects of exposure to carbon disulfide have been mainly studied among workers producing artificial fibres by the viscose method. It is possible that after severe chronic exposure organic deterioration occurs in the brain. The most noticeable findings are slowness in psychomotor performance and poor manual dexterity. Depression is also common (11, 34, 37).

## Carbon monoxide

Carbon monoxide is a common industrial poison which has neurotoxic effects. Exposure to it can be associated with heating ovens, central heating units, motor vehicle exhausts, foundries, and tobacco smoking. Its psychological effects are due to a lack of oxygen in the central nervous system. Common symptoms are unusual fatigue, lack of mental energy, irritability, and difficulty in concentrating. Even low levels of carboxyhaemo-globin (5%) have produced impairments in cognitive functions, and levels a little higher have affected manual dexterity and eye–hand coordination (1).

## Heavy metals

The most common neurotoxic heavy metals are lead and mercury. The main exposure sources of lead are mining, metal working, battery production, and paint production. Exposure to lead outside the place of work is also possible through food, water, and motor car exhausts. Among the subjective symptoms in workers chronically exposed to lead are unusual tiredness, gastrointestinal symptoms, and mood disturbances (15). After low-level, long-term exposure such functions as visuoconstructive ability, short-term memory, and manual dexterity diminished in correlation with the amount of lead in the blood (14).

Exposure to mercury can occur during metallurgic processing, in refineries, chloralkali plants, and certain chemical laboratories, and in the electrical industry. Its psychomotor and neuromuscular effects have been the subject of many studies, one of which was published in 1978 (20). The psychomotor functions have been shown to be sensitive to mercury and slight cognitive deterioration is also possible (6). Emotional alterations often lead to depression, listlessness, irritability, and social shyness.

## Pesticides

Certain pesticides, mainly the organophosphorus compounds, constitute one group of neurotoxic agents. They are used in agriculture and forestry. Acute intoxication usually gives rise to anxiety, dizziness, headache, and tremor. Depression is a common feature of chronic exposure and memory disturbance has also been observed (21, 33).

## Noise

Occupational exposure to noise is found among those working in the metal processing, lumber, and building industries, and those working in the midst of traffic. Noise is mainly generated by the machines and tools used for work. It can be continuous or intermittent (30). Its effects are related to its sound-pressure level, the duration and type of exposure, its temporal characteristics, and its frequency (2). They mainly depend on the physical features of the noise. A continuous, steady noise may have a very different effect from an impulsive noise. In some studies it was shown that unpredictable and complex impulsive noise can be more harmful to the hearing (3, 30) and more mentally stressful than continuous, steady noise (2, 9, 26).

Noise can influence man in many ways. It can damage the inner ear and affect vital life functions and wellbeing by causing distraction and disturbance. The effects on the hearing about which most is known are those caused by continuous, steady, broad-band noise and impulsive noise. Gun fire, which is an impulsive type of noise, can cause immediate and permanent damage to the hearing. The effects induced by noise can consist of either a temporal or a permanent shift in the threshold of hearing. The decibel limit for the risk of damage to the hearing is 85 dB(A) (17). In addition to affecting the hearing, noise can have effects that are mediated through the central and autonomic nervous systems (9). For example, it can effectively mask speech in the three octaves centred on 1 kHz, 2 kHz, and 4 kHz.

Continuous, steady noise can acutely improve performance and benefit the activation level of the human organism while a complex cognitive task requiring sustained vigilance is being performed (2, 9, 32). However, exposure to continuous, steady noise in monotonous types of work can cause the workers mental stress to the point that they become fatigued and have sleeping difficulties. In a visual task, vigilance and the speed of motor reactions can be slowed up and the responsiveness of the autonomic nervous system can be attenuated.

Impulsive noise can also cause stress to workers resulting in sleep disturbance and feelings of fatigue. However, their vigilance may be greater and their autonomic responsivity higher than those of workers exposed to continuous, steady noise.

Workers exposed to noise show more symptoms and annoyance than workers not exposed. Neuroticism and anxiety can be characteristics of those working in noise. Those exposed to impulsive noise can have difficulty in concentrating on their work and can feel that the noises distract them (28).

## Vibration

Workers who use machines and tools needing compressed air are subject to vibration, and noise can give rise to mechanical vibration (36, 39). Vibration can be sequential or involve the whole body. Hand-held mechanized tools generate sequential vibration. Whole-body vibration is involved when, for example, a worker does his job on or inside a machine. The frequency of whole-body vibration is 1–100 Hz and that of sequential vibration 1–2000 Hz.

In certain situations the body, or a part of the body, vibrates at the same frequency as another object. This causes some organs of the body to resonate in congruence with the other object; stomach and chest pains may result. The typical and early symptoms of vibration illness are aches and pains in the hands or the joints. Numbness of the hands, especially at night or after a working day, can also be a symptom of vibration illness (36, 41).

Because of the vasoconstriction caused by vibration the hands become less tolerant to cold and white fingers may result. The strength of the hand-grip can also decrease. All these symptoms can cause a decline in the motor performance needed for work. The head resonates at a frequency of 2 Hz; the eyes have been found to resonate at a frequency of 40–80 Hz. Resonance of the eyes can produce disturbances in visual function and in the coordination of the hands; this, in turn, can cause deterioration in performance based on visual and manual accuracy (*32, 36, 41*).

Vibration can affect the activation level of the human organism. Monotonous and continuous vibration can cause a decline in vigilance and activation, make the worker feel drowsy, and impair performance. Intermittent vibration, on the other hand, can have a stimulating and arousing effect, benefitting performance in a monotonous task requiring vigilance.

## Thermal conditions

The thermal conditions of a working environment are related to the climate and the location of the work (out of doors or indoors). The effects of thermal conditions during typical "hot" or "cold" occupations (e.g., baking and deep-freezing, respectively) depend on temperature, humidity, air velocity, the amount of radiant heat, the type of clothing used, the duration of exposure, the degree of physical activity performed during exposure and, of course, the health status of the exposed person.

Heat balance in the human body is maintained primarily by the circulatory system and through perspiring. Perspiring occurs when heat stress increases to the point that the circulatory system can no longer effectively regulate the heat balance of the body. If imbalance in the temperature of the body continues long enough, dysfunction in both the vital and the psychological functions results (*27, 32*).

Since, under heat stress, the circulation of blood in the body is directed towards the skin, the amount circulated to the muscles decreases. This phenomenon results in fatigue of the muscles and a decrease in their working capacity. There is a critical temperature, after which physical and cognitive performance is sharply impaired. Mental functions are also affected as a consequence of heat stress, primarily by changes in the activation level of the organism (*27, 32*). In many studies it has been shown that task performance varies at different temperatures. In the USA, national standards have been established in relation to work in hot environments which take mental performance in heat into account (40). In tasks requiring good concentration and clear thinking—involving reading and counting for example—performance may deteriorate in conditions of even moderate heat stress. In tasks demanding mechanical memory, recognition, and attention, however, performance can improve in conditions of heat stress that are slightly higher than moderate. In a task involving sustained attention and vigilance, too high a level of stress can impair vigilance and distort perceptual functions such as vision and hearing. The positive and negative effects of moderate levels of heat stress are likely to be greater for men than for women (42). Generally, men can tolerate higher temperatures better than women but women may be able to stand extreme conditions better (27).

## Improvement of physical and chemical environments

When improvements are being made in physical and chemical working environments, official or otherwise-accepted hygienic limits considered to be safe—e.g., the threshold limit value or the maximum acceptable concentration—are available as guidance. In setting the limits, very few data have been applied in relation to psychological effects—e.g., of chemical agents and physical agents—such as the extra-auditory effects of noise, one reason being that there are very few applied research data available in this field. In setting limits for such agents, the state of health of an average individual has usually been the starting point. They do not, therefore, protect every individual; sensitivity to adverse effects can vary according to sex, age, health status, or other predisposing characteristics.

Proper ventilation is important in conditions of exposure to a chemical agent. Other types of technical solution can be applied in conditions of exposure to a physical agent. Old work-places are usually difficult to redesign in a way that will protect the workers adequately. Another way of protecting them from adverse affects, although not a very good alternative, is to get them to use personal protective equipment—e.g., ear protectors against noise. Often, however, this is the only way to avoid high exposure.

## Education and research

When a worker starts a new job with the possibility of harmful exposure, adequate information and safety education must be provided from the outset. Knowledge of the correct working methods and information on the harmful effects of the agents he might encounter and the strategies to avoid them, motivate and help a worker to protect himself. Because of the varying sensitivity in individuals and because working conditions are not always such that harmful effects can be avoided, occupational health activities aimed at individuals or groups of workers are needed.

In the monitoring of mental health and of dysfunctions of the central nervous system, surveys of both subjective symptoms and psychological dysfunctions are possible. It is difficult, however, to monitor dysfunctions of the central nervous system. Data from the preexposure period would make intra-individual comparisons possible and increase the validity of the conclusions. Intra-individual comparison is often necessary because, for example, of the great variation in psychological functioning from individual to individual.

In both the Democratic Republic of Germany and Sweden, questionnaires on subjective symptoms were developed in order to monitor dysfunctions in the central nervous systems of workers exposed to neurotoxic agents (16, 35). The Swedish questionnaire concerns 16 symptoms usually associated with exposure to a solvent. Age-specific cut-off points are used to give an indication of the need for further clinical examination. It is proving to be a useful instrument as a first step in the screening of workers experiencing adverse effects due to chemical exposure.

A series of psychological tests is also a useful monitoring tool. Standardized series of tests for clinically examining workers exposed for long periods exist in the Democratic Republic of Germany, Finland, and Sweden. They are so time-consuming and difficult to interpret, however, that further refinement is needed before they can be used.

## References

1  BEARD, R. R. & GRANDSTAFF, N. Carbon monoxide exposure and cerebral functions. *Annals of the New York Academy of Sciences*, **174**: 385–395 (1970).

2  BROADBENT, D. E. Human performance and noise. In: Harris, C. M., ed. *Handbook of noise control*. New York, McGraw-Hill, 1979, pp. 17.1–17.20.

3  BURNS, W. *Noise and man, 2nd edition*. London, John Murray, 1973.

4  COLLIGAN, M. J. & SMITH, M. J. A methodological approach for evaluating outbreaks of mass psychogenic illness in industry. *Journal of occupational medicine*, **20**: 401–402 (1978).

5  ELOFSSON, S. A. ET AL. Exposure to organic solvents. A cross-sectional epidemiologic investigation on occupationally exposed car and industrial spray painters with special reference to the nervous system. *Scandinavian journal of work, environment and health*, **6**: 239–273 (1980).

6  FORZI, M. ET AL. Testpsychologische Leistungsfähigkeit in Quecksilberdampf-exponierten Arbeiten. [The effectiveness of psychological tests in the care of workers exposed to mercury vapour.] In: Klimkova-Deutschova, E. & Lukas, E., ed. *Proceedings of the 2nd International Industrial and Environmental Neurology Congress, Prague, September 1974*. Prague, Czechoslovak Neurological Society, 1976, pp. 70–73 (in German).

7  GAMBERALE, F. ET AL. The effect of styrene vapour on the reaction time of workers in the plastic boat industry. In: Horvath, M., ed. *Adverse effects of environmental chemicals and psychotrophic drugs*. Amsterdam, Elsevier, 1976, Vol. 2, pp. 135–148.

8  GRANDJEAN, E. ET AL. Investigations into the effects of exposure to trichloroethylene in mechanical engineering. *British journal of industrial medicine*, **12**: 131–142 (1955).

9  GULIAN, E. Noise as a stressing agent. *Psychologica*, **6**: 160–168 (1974).

10  HANE, M. ET AL. Psychological function changes among house painters. *Scandinavian journal of work, environment and health*, **3**: 91–99 (1977).

11  HÄNNINEN, H. Psychological picture of manifest and latent carbon disulphide poisoning. *British journal of industrial medicine*, **28**: 374–381 (1971).

12  HÄNNINEN, H. Psychological test methods: sensitivity to long-term chemical exposure. *Neurobehavioral toxicology and teratology*, **1**: 157–161 (1979).

13  HÄNNINEN, H. ET AL. Behavioral effects of long-term exposure to a mixture of organic solvents. *Scandinavian journal of work, environment and health*, **2**: 240–255 (1976).

14  HÄNNINEN, H. ET AL. Psychological performance of subjects with low exposure to lead. *Journal of occupational medicine*, **20**: 683–689 (1978).

15  HÄNNINEN, H. ET AL. Subjective symptoms in low-level exposure to lead. *Neurotoxicology*, **1**: 333–347 (1979).

16  HOGSTEDT, C. ET AL. A questionnaire approach to the monitoring of early disturbances in central nervous system. In: Aitio, A. et al., ed. *The biological monitoring of exposure to industrial chemicals*. Washington, DC, Hemisphere Publishing Corporation, 1982, pp. 275–287.

17  INTERNATIONAL ORGANIZATION FOR STANDARDIZATION. *Acoustics—assessment of occupational noise exposure for hearing conservation purposes*. Geneva, 1975 (ISO 1999).

18  KONIETZKO, H. ET AL. Zentralnervöse Schaden durch Trichloräthylen. [Damage to the central nervous system caused by trichlorethylene.] *Staub, Reinhaltung der Luft*, **35**: 240–241 (1975) (in German).

19  KOSKELA, R. A. ET AL. *Työntekijöiden mielipiteet työpaikkojen terveydellisistä oloista*. [*Opinions of workers on health conditions in their workplaces.*] Helsinki, Institute of Occupational Health, 1973 (Report No. 81) (in Finnish with summary in English).

20  LANGOLF, C. D. ET AL. Evaluation of workers exposed to elemental mercury using quantitative tests of tremor and neuromuscular functions. *American Industrial Hygiene Association journal*, **39**: 976–984 (1978).

21  LEVIN, H. S. & RODNITZKY, R. L. Behavioral effects of organophosphate pesticides in man. *Clinical toxicology*, **9**: 391–405 (1976).

22  LINDSTRÖM, K. ET AL. Disturbances in psychological functions of workers occupationally exposed to styrene. *Scandinavian journal of work, environment and health*, **3**: 129–139 (1976).

23  LINDSTRÖM, K. ET AL. Alcohol consumption and tolerance of workers exposed to styrene in relation to level of exposure and psychological symptoms and signs. *Scandinavian journal of work, environment and health*, **4**: Suppl. 2, 196–199 (1978).

24  LINDSTRÖM, K. & MARTELIN, T. Personality and long term exposure to organic solvents. *Neurobehavioral toxicology and teratology*, **2**: 89–100 (1980).

25  LINDSTRÖM, K. & SEPPÄLÄINEN, A. M. *Raudoittaja-ja maalaritutkimus: Osa 4. Korjausmaalarien oireet, psyykkiset suoritukset, neurofysiologiset löydökset sekä liuotinalitistus.* [*Study on reinforcers and painters: Part 4. Symptoms, psychological tests, neurophysiological findings and sensitization to solvents by correction painters.*] Helsinki, Institute of Occupational Health, 1980 (Report No. 170) (in Finnish with summary in English).

26  MCLEAN, E. K. & TARNOPOLSKY, A. Noise discomfort and mental health. *Psychological medicine*, **7**: 19–62 (1977).

27  MACKIE, R. R. & O'HANLON, J. F. A study of the combined effects of extended driving and heat stress on driver arousal and performance. In: Mackie, R. R., ed. *Vigilance: theory, operational performance, and physiological correlates.* New York, Plenum Press, 1977, pp. 537–558.

28  MANTYSALO, S. & VUORI, J. Visual reaction time, and autonomic responsivity after long-term exposure to impulse noise, and continuous noise (in preparation).

29  MARONI, M. ET AL. A clinical, neurophysiological and behavioral study of female workers exposed to 1,1,1-trichloroethane. *Scandinavian journal of work, environment and health*, **3**: 16–22 (1977).

30  MELNICK, W. Hearing loss from noise exposure. In: Harris, C. M., ed. *Handbook of noise control.* New York, McGraw-Hill, 1979, pp. 9.1.–9.16.

31  MÜNCHINGER, R. Der Nachweis Zentralnervöser Störungen bei lösungsmittel-exponierten Arbeiten. [Demonstration of disturbances in the central nervous system in workers exposed to solvents.] In: *Proceedings of the XIVth International Congress of Occupational Health, Madrid, 16–21 September 1963.* Amsterdam, Excerpta Medica Foundation, 1964, Vol. II, pp. 687–689 (in German, with an abstract in English in Vol. IV).

32  POULTON, E. C. Arousing stresses increase vigilance. In: Mackie, R. R., ed. *Vigilance: theory, operational performance, and physiological correlates.* New York, Plenum Press, 1977, pp. 437–453.

33  RODNITZKY, R. L. ET AL. Occupational exposure to organophosphate pesticides: a neurobehavioral study. *Archives of environmental health*, **30**: 98–103 (1975).

34  SCHNEIDER, H. Möglichkeiten der Psychodiagnostik bei neurotoxischen Expositionen. [The possibilities of psychodiagnosis in cases of exposure to neurotoxic materials.] In:

Horvath, M., ed. *Adverse effects of environmental chemicals and psychotropic drugs.* Amsterdam, Elsevier, 1976, Vol. 2, pp. 187–196 (in German).

35 SCHNEIDER, H. & SEEBER, A. Psychodiagnostik bei der Erfassung neurotoxischer Wirkungen chemischer Schadstoffe. [Psychodiagnosis in the assessment of the neurotoxic effects of noxious chemical substances.] *Zeitschrift für Psychologie,* **187**: 178–205 (1979) (in German).

36 STARCK, J. Finnish recommendation for maximum vibration levels. In: Korhonen, O., ed. *Vibration and work. Proceedings of the Finnish-Soviet-Scandinavian Vibration Symposium, Helsinki, 10–13 March 1975.* Helsinki, Institute of Occupational Health, 1975, pp. 117–120.

37 TUTTLE, T. C. ET AL. *Behavioral and neurological evaluation of workers exposed to carbon disulfide ($CS_2$).* Washington, DC, United States Government Printing Office 1977 (DHEW Publication No. (NIOSH) 77–128).

38 TUTTLE, T. C. ET AL. *A behavioral and neurological evaluation of dry cleaners exposed to perchlorethylene.* Washington, DC, United States Government Printing Office, 1977 (DHEW Publication No. (NIOSH) 77–214).

39 UNGAR, E. E. & COHEN, R. Vibration control techniques. In: Harris, C. M., ed. *Handbook of noise control.* New York, McGraw-Hill, 1979, pp. 20.1.–20.15.

40 UNITED STATES OF AMERICA. National Institute for Occupational Safety and Health. *Occupational exposure to hot environments: criteria for a recommended standard.* Washington, DC, United States Government Printing Office, 1973 (DHEW Publication No. (HSM) 73-10269).

41 WILLIAMS, N. Biological effects of segmental vibration. *Journal of occupational medicine,* **17**: 37–39 (1975).

42 WYON, D. P. ET AL. The mental performance of subjects clothed for comfort at two air temperatures. *Ergonomics,* **18**: 359–374 (1978).

# Part four

# Individual differences in susceptibility to stress

# Chapter 12

# Individual susceptibility and resistance to psychological stress

Richard S. Lazarus[1]

The common view of stress is that it is caused by a major catastrophic event, because that type of event is so damaging or disturbing to most people, psychologically and physiologically. That kind of thinking has led to an approach to stress measurement that involves an assessment of the number and severity of major life-events a person has experienced within a given period. In industrialized societies, the most severe of such events include loss of a loved one, loss of a job, or divorce. An alternative approach to stress measurement involves the assessment of *daily hassles* and *daily uplifts*. *Hassles* are the seemingly minor, annoying experiences of every-day living, such as traffic jams, difficult neighbours or co-workers, and too much or too little to do (*7, 13*). *Uplifts* are the positive counterparts of hassles, which produce pleasure or satisfaction—e.g., getting a good night's rest. These two approaches—assessment of major life-events and assessment of hassles and uplifts—are interwoven because, although most hassles arise independently of major life-events, the latter usually create new hassles and uplifts. It has been found that daily hassles are much better predictors of health status and morale than are life-events.

When stress is considered within an anthropological or sociological frame of reference—i.e., diverse societies or sub-groups are compared within a society—it becomes clear that the type and significance of both major and minor stressors, especially the latter, vary with culture, socioeconomic factors, and probably even with stage of life (*4, 12*). Such differences in reaction are the result of events having divergent meanings across social groups and cultures, as well as the chance and frequency with which they occur. This divergency in meaning applies less to major life-events than to hassles and uplifts, since their significance is more subtle and dependent on cultural values.

The sociocultural variation referred to above is comparable, in a sense, to the ubiquitous individual differences in all the variables and processes of stress. Just as differences between groups may be ascribed to different cultural values and social structures, so can differences between individuals arise because they vary greatly in their life-styles and personality traits. The

[1] Department of Psychology, University of California, Berkeley, California, United States of America.

differences account for the various ways in which individuals evaluate and cope with an encounter.

Theoretical work has concentrated, to a large extent, on cognitive appraisal and coping (16, 17, 19, 20). Psychological stress depends on how a person appraises an encounter with respect to its implications for his own wellbeing. This involves a perception, conscious or unconscious, of whether he will be affected by the outcome and whether he has sufficient resources to cope with it. The influence of the encounter on his feelings, the way he reacts socially and at work, and his somatic health cannot be understood by reference to the stressor alone; it depends on his own cognitive appraisal and coping functions, which are in constant flux as an encounter unfolds.

Two concepts, representing opposite sides of the same coin, express this idea of individual differences in patterns of stress, coping, and adaptation, namely, *susceptibility* and *resistance*.

## Susceptibility

Susceptibility—or vulnerability—refers to one individual's tendency to react to certain types of encounter or situation with psychological stress, or with a greater degree of stress than another individual. When an environment places severe demands on a group of individuals, as in the event of natural and man-made disasters, such as famine, war, or disabling accidents—all major life-events—most of them will react with psychological and physiological disturbances, though the degree of disturbance and the ways in which they struggle to cope show great variability (3). Because of the relative universality of the reactions, such extreme environmental conditions usually warrant being called stressors. However, many routine conditions of life, including those in the working context, are not stressors for most individuals, yet they generate disturbance in some. They may include such routine parts of life as meeting others, being evaluated, relating in an affectionate way, managing ordinary social and working responsibilities, and utilizing leisure time.

The principle of suceptibility can be illustrated with observations from findings relating to hassles and uplifts in daily life. It must not be imagined that the frequency with which hassles occur or their intensity, provide simple measures of stressful occurrences, as is the case in the simple atheoretical input-output tradition of life-events research. When a person attests to a hassle it reflects how he is subjectively experiencing and appraising the transactions of living. When he reports that he has had a specific encounter—e.g., been caught in a traffic jam—he is referring to an actual, objective, experience but the fact that he has attested to being hassled (or uplifted by another type of encounter) means that the encounter has been of personal significance, and that is what makes it of importance.

Thus, in what seems like a paradox, chronically ill people report uplifts, such as getting a good night's sleep or feeling energetic, more readily than well people; presumably because a sick person is more likely to notice, and be grateful for, such positive experiences than a healthy person, who takes them for granted. The same principle applies to hassles; their psychological significance depends on the person's conditions of life. For example, a person might be more likely to attest to inclement weather as a hassle when he is on vacation in August than when he is back in the office in September (6, 11).

So far very little attention, in research on stress, has been given to factors in a person's life and personality that influence differential susceptibility to feeling harmed, threatened, or challenged—three stress appraisals with differing adaptational consequences. It is important to endeavour to discover what makes one person more, or less, susceptible than another to stress when he engages in certain kinds of transaction with the environment. Two personality variables appear to be particularly promising as predictors of susceptibility to psychological stress: an individual's distinctive pattern of commitments and his beliefs about himself and the world (18). These are by no means the only variables but they both offer interesting possibilities for adding to the already substantial number of research and theoretical findings.

Commitments are an expression of a person's ideals and goals and the choices he is willing to make to realize those ideals and goals; psychologists have traditionally placed them under the rubric of *motivation*. Some people share some commitments because of their common biological or social heritage but patterns of commitment also vary greatly. For some, achievement is the most powerful motivator; for others, achievement is of minimal importance compared with the quality of their relationships with others. The degrees to which commitments are important affect susceptibility to stress because encounters that endanger strong commitments are more likely to lead to threat or harm appraisals than those that endanger weak commitments (22, 31).

The manner in which commitments affect susceptibility to stress is complicated, in that a commitment can serve not only as a vulnerability but as a resource, insofar as it protects against boredom, meaninglessness, and alienation, the psychological disorders of affluence. In a study of business executives published in 1979 it was suggested that a committed person is less likely to suffer stress-induced disease (*14*). It was postulated that so-called *hardy* persons—i.e., those who, under stress, do not succumb to disease—are those who, in addition to holding strong commitments, believe that they can control or influence events and regard changes in their lives as challenges rather than threats. This type of research, addressing personality variables that might be expected to affect either susceptibility or resistance to stress and their probable health consequences, is rare. It is unfortunate, therefore, that methodological difficulties weaken the empirical evidence for the appealing argument put forth. For example, the reliance that can be placed on the interpretations from the 1979 study is limited by the fact that the antecedent personality variables contributing to hardiness could be confounded with health outcome variables; the construct validity of the conceptual labels used to determine the personality traits comprising hardiness also presents problems. Finally, inferences as to how hardy people react to, and cope with, stressful encounters are not verified by direct observation; they remain inferences from questionable antecedent trait measures.

There is much interest in the role a person's *beliefs* play in his capacity to *control* the outcome of encounters (*8, 10, 15, 21, 23, 25, 28, 32*). A review of research on locus of control (*24*) based on work published in 1966 (*26*) indicates that the tendency has been to confound two components of control distinguished by Bandura (*2*): *efficacy expectations*, a person's conviction that he can or cannot perform the acts necessary for effective coping; and *outcome expectations*, the conviction that the environment will or will not be responsive to competent efforts to change it. It was implied that a negative belief in regard to either of those components will increase the likelihood of psychological stress. In other words, a poor sense of self-efficacy, whether or not the self-assessment is accurate, increases susceptibility to stress in any context in which it is relevant.

## Resistance

On the other side of the coin to susceptibility is *resistance* to stress; susceptibility and resistance can be regarded as forming a dimension in any given area of experience. The presence of *resistance resources* should decrease stress insofar as they facilitate the management of potentially stressful encounters (*1*). Such resources include knowledge, skills, financial security, access to social support, and living in a social environment of awareness of being accepted, valued, or loved (*5*). The term *social support* must not be used casually, however, without careful examination of the support proferred, since it can have a negative, as well as a positive, effect on stress, coping, and adaptation (*27, 29, 30*).

Since stress is inevitable, of the major contributors to individual differences in susceptibility or resistance to it, none are more important or less well understood than the factors involved in how a person copes with the demands of living. A prerequisite for healthy functioning is not only an ability to master the environment but the wisdom to come to terms with conditions that cannot be changed. For that reason, an assessment of emotion-focused coping, as well as problem-focused coping, is important in studying how a person manages a stressful encounter (*9*). In a society such as that of the USA with its cultural value of mastery rather than acceptance problem-focused coping is more readily appreciated. When mastery is not possible or wise, however, emotion-focused coping allows a person to tolerate or accept harsh circumstances, or to reinterpret them in a less threatening way.

Research findings strongly suggest that both kinds of coping are to be found in most stressful encounters, albeit with great individual differences, in emphasis on one or on the other, and in the particular emotion-focused modes employed, such as denial, avoidance, intellectualized detachment, and externalization of blame. Little is known in regard to the optimum proportions of the two kinds of coping necessary for successful adaptation but it is suspected that both are essential to healthy functioning. Through theory and research, the primary task is to determine the conditions in which particular coping patterns will result in a desirable or an undesirable outcome. This will contribute to an understanding of successes and failures in adapting.

Since people vary in their susceptibility to stress, it is important to study the factors contributing to coping patterns, at work and in other settings, in order to gain the knowledge that will enable the settings to be changed so that good coping is made easy in the interest of productivity, health, and wellbeing.

# References

1 ANTONOVSKY, A. *Health, stress and coping: new perspectives on mental and physical well-being.* San Francisco, CA, Jossey-Bass, 1979.

2 BANDURA, A. Self-efficacy: toward a unifying theory of behavioral change. *Psychological review,* **84**: 191–215 (1977).

3 BENNER, P. ET AL. Stress and coping under extreme conditions. In: Dimsdale, J. E., ed. *Survivors, victims, and perpetrators: essays on the Nazi holocaust.* Washington, DC, Hemisphere, 1980, pp. 219–268.

4 BRIM, O. G., JR & RYFF, C. D. On the properties of life events. In: Baltes, P. B. & Brim, O. G., Jr, ed. *Life-span development and behavior.* New York, Academic Press, 1980, Vol. 3, pp. 367–388.

5 COBB, S. Social support as a moderator of life stress. *Psychosomatic medicine,* **38**: 300–314 (1976).

6 COSTA, P. T. & McCRAE, R. R. Somatic complaints in males as a function of age and neuroticism: a longitudinal analysis. *Behavioral medicine,* **3**: 345–358 (1980).

7 DELONGIS, A. ET AL. Relationship of daily hassles, uplifts, and major life events to health status. *Health psychology,* **1**: 119–136 (1982).

8 FOLKMAN, S. Personal control and stress and coping processes: a theoretical analysis. *Journal of personality and social psychology,* **46**: 839–852 (1984).

9 FOLKMAN, S. & LAZARUS, R. S. An analysis of coping in a middle-aged community sample. *Journal of health and social behavior,* **21**: 219–239 (1980).

10 GARBER, J. & SELIGMAN, M. E. P. *Human helplessness: theory and applications.* New York, Academic Press, 1980.

11 HELSON, H. Adaptation level theory. In: Koch, S., ed. *Psychology: a study of a science.* New York, McGraw-Hill, 1959, Vol. 1.

12 HULTSCH, D. F. & PLEMONS, J. K. Life events and life-span development. In: Baltes, P. B. & Brim, O. G., Jr. ed. *Life-span development and behavior.* New York, Academic Press, 1979, Vol. 2, pp. 1–36.

13 KANNER, A. D. ET AL. Comparison of two modes of stress measurement: daily hassles and uplifts versus major life events. *Journal of behavioral medicine,* **4**: 1–39 (1981).

14 KOBASA, S. C. Stressful life events, personality, and health: an inquiry into hardiness. *Journal of personality and social psychology,* **37**: 1–11 (1979).

15 LANGER, E. J. The illusion of control. *Journal of personality and social psychology,* **32**: 311–328 (1975).

16 LAZARUS, R. S. *Psychological stress and the coping process.* New York, Toronto, and London, McGraw-Hill, 1966.

17 LAZARUS, R. S. The stress and coping paradigm. In: Eisdorfer, C. et al., ed. *Models for clinical psychopathology.* New York, Spectrum, 1981, pp. 177–214.

18 LAZARUS, R. S. & DELONGIS, A. Psychological stress and coping in aging. *American journal of psychology,* **38**: 245–254 (1983).

19 LAZARUS, R. S. & FOLKMAN, S. Coping and adaptation. In: Gentry, W. D., ed. *The handbook of behavioural medicine.* New York, Guilford Press, 1984, pp. 282–325.

20 LAZARUS, R. S. & LAUNIER, R. Stress-related transactions between person and environment. In: Pervin, L. A. & Lewis, M., ed. *Perspectives in interactional psychology.* New York, Plenum Press, 1978, pp. 287–327.

21 LEFCOURT, H. M. *Locus of control: current trends in theory and research,* 2nd edition. Hillsdale, NJ, Erlbaum, 1982.

22 MAHL, G. F. Physiological changes during chronic fear. *Annals of the New York Academy of Sciences,* **56**: 240–249 (1953).

23 MILLER, S. M. When is a little information a dangerous thing? Coping with stressful events by monitoring versus blunting. In: Levine, S. & Ursin, H., ed. *Coping and health.* New York, Plenum Press, 1980, pp. 145–169.

24 PETERSON, C. *The sense of control over one's life: a review of recent literature.* Unpublished.

25 ROTHBAUM, F. ET AL. Changing the world and changing the self: a two-process model of perceived control. *Journal of personality and social psychology,* **42**: 5–37 (1982).

26 ROTTER, J. B. Generalized expectancies for internal versus external control of reinforcement. *Psychology monographs,* **80**: 1–28 (1966).

27 SCHAEFER, C. ET AL. The health-related functions of social support. *Journal of behavioral medicine,* **4**: 381–406 (1982).

28 SILVER, R. L. & WORTMAN, C. B. *Expectations of control and coping with permanent paralysis. Paper presented at the Symposium on Issues of Control in Health, 88th Annual Convention of the American Psychological Association, Montreal, September 1980.* Washington, DC, American Psychological Association, 1980.

29 SULS, J. Social support, interpersonal relations, and health: benefits and liabilities. In: Sanders, G. S. & Suls, J., ed. *Social psychology of health and illness.* Hillsdale, NJ, Erlbaum, 1982.

30 THOITS, P. A. Conceptual, methodological, and theoretical problems in studying social support as a buffer against life stress. *Journal of health and social behavior,* **23**: 145–159 (1982).

31 VOGEL, W. ET AL. Intrinsic motivation and psychological stress. *Journal of abnormal and social psychology,* **58**: 225–233 (1959).

32 WORTMAN, C. B. & BREHM, J. W. Responses to uncontrollable outcomes: an integration of reactance theory and the learned helplessness model. In: Berkowitz, L., ed. *Advances in experimental social psychology.* New York, Academic Press, 1975, Vol. 8, pp. 227–336.

# Chapter 13

# Stress-prone behaviour:
# Type A pattern

Cary L. Cooper[1]

In the early 1960s a specific approach was adopted in research on individual differences in reactions to stress, the results of which showed a relationship between behavioural patterns and the prevalence of coronary heart disease (8, 21, 22). It was found that individuals manifesting certain behavioural traits were significantly more at risk. Such individuals came to be referred to as exhibiting the *coronary-prone behaviour pattern Type A*, as distinct from *Type B*—the pattern exhibited by those at low risk to coronary heart disease. The Type A pattern was described as an overt behavioural syndrome or type of living, characterized by extremes of competitiveness, striving for achievement, aggressiveness, haste, impatience, restlessness, hyperalertness, explosiveness of speech, tenseness of the facial musculature, and feelings of being under pressure of time and the challenge of responsibility. It was suggested that people with a Type A pattern of behaviour are often so deeply involved and committed to their work that other aspects of their lives are relatively neglected (*13*).

In the early studies individuals were designated as Type A or Type B on the clinical judgements of physicians and psychologists or as a result of peer ratings; a higher incidence of coronary heart disease was found among Type A individuals than among Type B individuals. There were, however, inherent methodological weaknesses in this approach which were overcome during what has come to be recognized as the classic study, initiated in 1960 (*21, 22*). This, called the Western Collaborative Group Study, was a prospective study (unlike the earlier studies which were retrospective) of a sample of 3182 men apparently free of coronary heart disease at the outset of the study. All the men were rated either Type A or Type B after intensive interviews by psychiatrists, who did not have access to their biological data, and without having been seen by a cardiologist. The diagnoses were made by an electrocardiographer and an independent physician, who were not informed of the individuals' behavioural patterns. After $2\frac{1}{2}$ years it was found that the Type A men aged 39–49 years had 6.5 times the incidence of coronary heart disease as the Type B men, and those aged 50–59 years had 1.9 times the incidence. They also demonstrated risk factors through elevated serum cholesterol levels, elevated beta-lipoproteins, decreased blood clotting times,

[1] Department of Management Sciences, University of Manchester, Manchester, England.

and elevated daytime excretions of norepinephrine. After $4\frac{1}{2}$ years the relationship of Type A behavioural pattern to incidence of coronary heart disease remained the same. In terms of the clinical manifestations of coronary heart disease, among those exhibiting Type A behavioural patterns the incidence of acute myocardial infarction and angina pectoris was significantly higher; the incidence of clinically unrecognized myocardial infarction was also higher (21, 22). It was also found that the risk of recurrent and fatal myocardial infarction was significantly related to Type A characteristics (20).

The same results were obtained in a study of Trappist and Benedictine monks (17). The Type A coronary-prone group—determined by a double-blind procedure—showed 2.3 times the incidence of angina pectoris and 4.3 times the incidence of myocardial infarction as the group judged to be Type B. Many other studies have shown roughly the same results (1, 24).

## Type A behaviour and the working environment

Much of the research into Type A behaviour in the working environment suggests that the stressors within the environment itself enhance Type A behaviour patterns. It was postulated that often an individual—e.g., a switchboard operator, a taxi-cab driver, an assembly-line worker, or a manager—does not possess a Type A pattern of behaviour on entering an occupation (19). However, time pressures and the conscientiousness demanded by the job can make a relaxed Type B into a Type A, or a less extreme Type A into a more exaggerated Type A.

Recent research on the effects of stressors on managers substantiates this assertion to some degree suggesting that the conditions most responsible for facilitating Type A behaviour are those encountered in the working environment (12). Certainly an earlier physiological study of a group of managers supports the suggestion (10); Type A middle-aged managers were found to have higher catecholamine excretion rates and were more physiologically aroused during working hours than the Type B individuals. Occupational stressors can have similar physiological effects on Type B individuals. It was discovered, for example, that Type B air traffic controllers have a higher incidence of hypertension than their Type A colleagues; they blamed the chronic and unpredictable stressors at work for their poor health (18).

When examining Type A behaviour patterns in the working environment, the contention that an important facet of Type A behaviour is a preference to control the environment is worthy of note (3). Consequently, the most serious stressors in a working situation are those the individual perceives might lead to decreased levels of control. The factors in the working environment that can give rise to this category of stressor include job involvement, responsibility for people and things, role ambiguity, role conflict, promotion in excess of capability, lack of participation in decision-making, poor working relationships, and work overload (2, 3, 4, 12, 16). A number of studies have concentrated on investigating the relationship between Type A behaviour and work overload, working relationships, job involvement, and job satisfaction (3).

## Work overload

In several studies strong correlations have been found between work-load and anxiety in Type A individuals in varying occupations—e.g., university staff working on computers (2), male accountants (9), and managers (3). However, an important question raised is how much is work overload self-imposed by the Type A individual (7).

## Working relationships

Dissatisfaction with subordinates and feelings of being misunderstood by supervisors have been voiced by workers with Type A patterns of behaviour; in some cases there is evidence that the dissatisfaction may be partially self-imposed (6, 7, 12). It was pointed out that, compared with Type B individuals, Type A individuals are more inclined to wish to work alone when subjected to acute stress and this enables them to give themselves deadlines and increase their work-loads (7). Thus, Type A individuals increase their own stress by reducing the opportunity to receive support from co-workers and subordinates, and this may, in turn, enhance their feelings of frustration towards their colleagues.

## Job satisfaction

It has been shown that there is a correlation between coronary heart disease and job dissatisfaction (11, 23); consequently a similar correlation between a Type A behaviour pattern and job dissatisfaction might be expected. However, this does not seem to be the case. In one recent study no significant differences were found in the levels of job satisfaction of Type A managers and Type B managers but Type A managers could be associated with a higher company growth rate (*12*).

## Job involvement

Deeper job involvement, enhanced by a highly developed need to achieve, has been associated with Type A behaviour (*16*). This seems to be compounded by the additional characteristics of competitiveness and high self-expectations in regard to work performance. Therefore, although a number of high-risk working environment variables have been determined, the self-selection of Type A individuals into work settings conducive to Type A behaviour is an issue that requires further research (*3*). In fact, it is maintained by certain experts that Type A individuals possess behavioural traits that actually facilitate self-selection into occupations involving increased exposure to stressors (*14*).

## Fit between individuals and organizations

A recent study was based on the proposition that organizations, as well as people, can be classified as having a Type A or Type B pattern of working environment and that the resulting match, or lack thereof, is related to various health indices (*15*). It was found that among 315 medical technologists: (1) Type B individuals working in Type B organizations report the fewest symptoms of ill-health; (2) Type A individuals working in Type A organizations report the most symptoms of ill-health; and (3) Type B individuals working in Type A organizations and Type A individuals working in Type B organizations report an intermediate level of symptoms of ill-health.

Extensive research reviews, such as those of Davidson & Cooper (*5*) and Chesney & Rosenman (*3*) provide further information on Type A behaviour.

## References

1  BORTNER, R. W. & ROSENMAN, R. H. The measurement of pattern A behavior. *Journal of chronic diseases*, **20**: 525–533 (1967).

2  CAPLAN, R. D. & JONES, K. W. Effects of workload, role ambiguity and Type A personality on anxiety, depression and heart rate. *Journal of applied psychology*, **60**: 713–719 (1975).

3 CHESNEY, M. A. & ROSENMAN, R. H. Type A behaviour in the work setting. Cooper, C. L. & Payne, R., ed. *Current concerns in occupational stress.* Chichester, New York, Brisbane, and Toronto, Wiley, 1980, pp. 187–212.

4 CUMMINGS, T. G. & COOPER, C. L. A cybernetic framework for studying occupational stress. *Human relations,* **32**: 395–418 (1979).

5 DAVIDSON, M. J. & COOPER, C. L. Type A coronary-prone behavior in the work environment. *Journal of occupational medicine,* **22**: 375–383 (1980).

6 DAVIDSON, M. J. & VENO, A. Stress and the policeman. In: Cooper, C. L. & Marshall, J., ed. *White collar and professional stress.* Chichester, New York, Brisbane, and Toronto, Wiley, 1980, pp. 131–166.

7 DEMBROSKI, T. M. ET AL. Components of the psychomotor performance challenge. *Journal of behavioral medicine,* **1**: 159–176 (1978).

8 FRIEDMAN, M. *Pathogenesis of coronary artery disease.* New York, McGraw Hill, 1969.

9 FRIEDMAN, M. ET AL. Plasma catecholamine response of coronary-prone subjects (Type A) to a specific challenge. *Metabolism,* **24**: 205–210 (1975).

10 FRIEDMAN, M. ET AL. Excretion of catecholamines, 17-ketosteroids, 17-hydroxy-corticoids and 5-hydroxyindole in men exhibiting a particular behavior pattern (A) associated with high incidence of clinical coronary artery disease. *Journal of clinical investigation,* **39**: 758–764 (1960).

11 HOUSE, J. S. Occupational stress and coronary heart disease: a review and theoretical integration. *Journal of health and social behavior,* **15**: 12–27 (1974).

12 HOWARD, J. H. ET AL. Work patterns associated with Type A behavior. *Human relations,* **30**: 825–836 (1977).

13 JENKINS, C. D. Psychologic and social precursors of coronary disease. *New England journal of medicine,* **284**: 244–255 (1971).

14 McMICHAEL, A. J. Personality, behavioural, and situational modifiers of work stressors. In: Cooper, C. L. & Payne, R., ed. *Stress at work.* Chichester, New York, Brisbane, and Toronto, Wiley, 1978, pp. 127–147.

15 MATTESON, M. T. & IVANCEVICH, J. M. Type A and B behavior patterns and self-reported health symptoms and stress: examining individual and organizational fit. *Journal of occupational medicine,* **24**: 585–589 (1982).

16 MATTHEWS, K. A. & SAAL, F. E. Relationship of the Type A coronary-prone behavior pattern to achievement, power and affiliation motives. *Psychosomatic medicine,* **40**: 631–637 (1978).

17 QUINLAN, C. B. ET AL. *The association of risk factors and coronary heart disease in Trappist and Benedictine monks. Paper presented at a meeting of the American Heart Association, New Orleans, Louisiana, 1969.* Dallas, TX, American Heart Assocation, 1969.

18 ROSE, R. M. ET AL. *Air traffic controller health change study: a prospective investigation of physical, psychological and work-related changes.* Springfield, VA, United States National Technical Information Service, 1978 (Office of Aviation Medicine Report No. FAA-AM-78-39).

19  Rosenman, R. H. Type A behavior and ischemic heart disease. *Karger gazette*, **37**: 1–8 (1978).

20  Rosenman, R. H. et al. Clinically unrecognized myocardial infarction in the Western Collaborative Group Study. *American journal of cardiology*, **19**: 776–783 (1967).

21  Rosenman, R. H. et al. A predictive study of coronary heart disease: the Western Collaborative Group Study. *Journal of the American Medical Association*, **189**: 15–26 (1964).

22  Rosenman, R. H. et al. Coronary heart disease in the Western Collaborative Group Study. *Journal of the American Medical Association*. **195**: 86–92 (1966)

23  Sales, S. M. & House, J. S. Job satisfaction as a possible risk factor in coronary disease. *Journal of chronic diseases*, **23**: 861–873 (1971).

24  Zyzanski, S. J. & Jenkins, C. D. Basic dimension within the coronary-prone behavior pattern. *Journal of chronic diseases*, **22**: 781–795 (1970).

Chapter 14

# Age and gender in relation to stress at work

1. Hadžiolova[1]

## Introduction

From a review of the large number of publications on aging that have accumulated over the past 20 years it is clear that a decreased ability to adapt to environmental challenges is a fundamental manifestation of growing old (20).

All stages of aging are characterized by two contradictory phenomena: (a) a deterioration in bodily structure and functions; and (b) a mobilizing process that comprises several adaptive and regulatory mechanisms aimed at preserving viability and increasing life expectancy (13). This means that changes during a lifespan in responses to stress are complex. Reaction to stress at different ages can be assessed through the physiological responses—endocrine, cardiovascular, and respiratory—and the behavioural responses—lowered performance rate, increase in errors, fatigue, impaired coordination, and changed emotional activity.

Despite the importance of age as a determinant in reactions to stress, its parameters and mechanisms are not sufficiently understood. Those considered here will relate to: age-related changes in the hypothalamic pituitary-adrenal-cortical and sympathetic-adrenal-medullary systems; changes with age in the responses of the two systems to occupational stressors; and variations with age in how stressful events at work are perceived and the importance given to them.

The effect of gender on responses to stressors will be discussed separately, though its complexity and the discrepancies and gaps in knowledge make a comprehensive study impossible.

## Age and stress at work

### Neuroendocrine changes and aging

It was observed by several authors writing on endocrinology and aging that changes in the reaction to stress with age depend on the effect of age on the structure and hormonal content of the glands, the response of the glands to stimuli, the kinetics of hormonal distribution and metabolism, and the sensitivity of the organs concerned to the hormones (1, 8, 14, 15).

[1] Institute of Hygiene and Occupational Health, Sofia, Bulgaria.

Hypothalamic aging is being studied extensively because of its essential role in the entire aging process and its basic importance for homeostasis. A hallmark of aging is that hypothalamic activity, and thus the physiological systems, are disturbed, the disturbances appearing as alterations in a person's ability to adapt to his environment (13). Aging causes significant changes in the metabolism of neurotransmitters, with consequent effects on the production of releasing hormones. Catecholamine functions, particularly dopamine pathways, in the hypothalamus are highly vulnerable to aging (9).

The amount of adrenocorticotropic hormone (ACTH) in the pituitary gland does not appear to change with age and it has been observed that the functional state of the adenohypophysis is not impaired in old age (5). No differences between elderly and young subjects were found in the levels or the circadian variations of serum ACTH.

One of the basic parameters for assessing stress reaction is the plasma cortisol level. It is therefore necessary to know how that hormone's basal level, which is overlapped by stress-induced changes, fluctuates. Despite some differences in investigation results, the prevailing opinion is that neither basal plasma cortisol levels nor cortisol circadian rhythm are subject to age-related change. It has been shown that in women of up to 60 years age does not have a statistically significant influence on cortisol circadian rhythm (27). However, plasma cortisol level is not only determined by secretion rate, it is also influenced by the kinetics of the hormone's distribution and metabolism. In the elderly the 24-hour cortisol secretion rate is 75% that of younger people and there is a significant decrease in the exchange of corticosteroids between the blood and the tissues. However, the slower rate of steroid metabolism in the elderly may be due to such other factors as a decline in their utilization by target tissues, changes in their binding to plasma proteins, and decreases in liver metabolism and the clearance of metabolites from the plasma (35).

A brief review of the literature shows how complex are the changes in adrenocortical activity with age. Glucocorticoid homeostasis is maintained by a reduction in the disposal of corticoids and in the sensitivity of the adrenal glands to ACTH.

Steroid products excreted in the urine—such as 11-oxycortico-steroids (11-OCS), 17-oxycorticosteroids (17-OCS), and 17-keto-steroids (17-KS) can be easily measured and have long been accepted as indicators of stress at work. Hence, data on age-related changes in the basal levels of excreted steroids are of particular interest. Numerous studies have shown that the excretion of 17-KS is significantly modified with age. The highest values are recorded in men aged between 20 and 30 years, after which a gradual decrease is observed with the tendency particularly evident after 40 years of age. In women 17-KS values are generally lower and age-related changes less evident. The excretory levels of 17-KS in men in their 50s change much more than in women at the same age; later the values decrease steadily in both sexes. Thus, aging is associated with a gradual fading of adrenal cortex androgenic activity, the tendency being more evident in men (22, 23, 34, 37). Changes in the excretion rates of 17-OCS in the urine are less marked than in the excretion rates of 17-KS. Some investigators found a lowered excretion of 17-OCS after the age of 50 years and others only after the age of 60 years.

Enhanced sympathetic-adrenal-medullary activity is an important indicator in human stress reaction and information on age-related changes in plasma and urine catecholamine secretion rates in man is particularly valuable. Modern assay techniques now allow age-related modifications to be determined in plasma catecholamine secretion rates. In a study of subjects aged 35–65 years, for example, a linear relationship between age and plasma noradrenaline was established, while they were recumbent and while they were in an upright position. The plasma noradrenaline secretion rate also increased in older persons when they were standing up (32). In another study, adrenaline secretion rates were not affected by age but noradrenaline secretion rates in older persons were 28% greater during the day and 75% greater at night (29). The data on age-related changes in urine catecholamine excretion rates while the subject is at rest are not so consistent. One investigator could not establish any differences in catecholamine excretion rates in subjects aged 17–29 and 30–59 years (21). Others could find no age- or gender-related differences in subjects aged up to 50 years, though increased rates were found in women aged 51–60 years (23). In studying the circadian rhythm of catecholamine secretion in subjects aged 20–30 and 31–65 years, no significant differences in the amplitudes and acrophases of fluctuations in excretion rates were found between the two age groups. A certain decrease of noradrenaline and adrenaline was observed in the second group, the changes being more evident for adrenaline (17).

Despite the obvious need for further, more detailed, studies, it can be deduced from the data available that significant changes in basal urine catecholamine levels are not observed at under 60 years of age. In conclusion, the data presented above on changes in the basal levels of activity of the hypothalamic pituitary-adrenal-cortical and sympathetic-adrenal-medullary systems do not provide evidence of age-related functional deficiencies when the subjects are resting. Nevertheless, a number of differences between younger and elderly subjects in the magnitude of the responses to stressors and the rates of recovery after stress were demonstrated.

## Changes with age in the responses of the systems to occupational stressors

The assessment of changes with age in responses to stress presents considerable difficulties. Currently available data are mainly related to subjects past the age of 60–65 years, whose responses to stimuli are clearly different from those of younger subjects, and who are characterized by a decrease in response and in the rate of recovery after stress. The data, which derive mainly from plasma 17-OCS and urine steroid metabolite levels, are of uncontestable value in gerontology but are little used in occupational psychophysiology. Much of the data are derived from studies on subjects of varying ages, performing different tasks in stressful situations under experimental conditions. The results are of prime importance, making it possible to define the factors that influence the systems precisely and to establish the age-related differences. It has to be recognized that to carry out this type of research under actual working conditions is extremely difficult and there are few field studies.

Intense physical load at work is an important stress factor; the subjective evaluation of the intensity of physical load is known to change with age (4). In men, subjective evaluation of the intensity of a load corresponds with pulse rate; in women the subjective evaluation anticipates the intensity of the load. Changes in steroid excretion rates correspond to the subjective evaluation. Adrenal-cortical activity is exhausted earlier in older subjects in spite of an initially greater increase in the levels of 17-OCS and 17-KS (33).

A number of investigators who gave particular attention to changes in plasma and urine catecholamine secretion rates under conditions of stress in subjects of varying ages, noted the higher activity of the sympathetic nervous system in older people. Older people with high basal plasma noradrenaline secretion rates show a greater increase in such rates after physical exercise (38). Older people may display a marked response to certain other stressors;

in response to cold, for example, people over 40 years of age secreted more noradrenaline and had higher blood pressure than younger people (3, 28). No difference was noticed in plasma adrenaline secretion rates whether the subject was resting or experiencing stress and in both groups pulse rate and blood pressure increased, the difference in heart rate being more evident in the younger people (3).

When studying stressful situations at work, investigators concentrated on either young people who had just started to work, or on those at the other extreme of the age spectrum—near or after retirement. A subject's ability to adapt to the stress of retirement is conditioned, among other things, by the type of work he did before, which defines to a great extent his health status, social resources, and satisfaction with life. Thus, overall adaptation to old age could be regarded as a continuation of life events and individual patterns of adaptation (18).

The interrelationship of stress, health, and coping strategies throughout life has been a matter for investigation. It was found that coping strategies remain unchanged with advancing age (37). They are largely dependent on health conditions. Older subjects intentionally avoid potentially stressful situations, thus modifying their response (24).

When studying stress among industrial workers, defining the boundary beyond which a worker is considered to be "elderly" presents an additional difficulty. This critical boundary may fluctuate considerably because of the number of determining factors involved, such as the degree and character of the physical and mental work-load, the working conditions, possibilities for controlling the work-load, the rapid progress of technology, and the distribution of labour. Other contributing factors include the worker's general education and special training, his marital status, and his state of health (25).

Specific occupational factors that could cause overloading and stress in older workers are: working under pressure of time and frequent changes in the technological process; accelerated working pace and the need to process large amounts of information; production line and piece work; work that requires a high level of vigilance; hard manual labour; and unfavourable working conditions, such as dust and noise. Numerous investigations performed in the Federal Republic of Germany of occupational stress among older workers have shown that, with advancing age, industrial workers complain of strain at work to an increasing extent. For example, in one study only 10% of young workers considered themselves overloaded and under strain, while 19% of the workers over 50 years complained in such a way (25).

Shift work is considered to be another important occupational stressor, especially for older workers, for whom it becomes increasingly difficult to adjust (36). The daytime sleep of the older worker (50–59 years of age) is more disturbed and restless than that of the younger worker, and he has higher noradrenaline excretion and diuresis rates when he sleeps during the day. The negative effects of night work can thus be considered to be exacerbated by increasing age.

Nevertheless, when studying the adaptive potential of the older worker a number of inherent positive characteristics must be given due regard, such as his greater professional experience and knowledge, sense of responsibility, and ability to resolve problems. These qualities can compensate to a considerable extent for any decrease in working capacity. That is why in a number of professions elderly workers can continue their professional duties without additional strain or overloading (26).

Teachers aged 25–55 years, assessed during their routine work, did not show any significant age-related differences in adrenaline, noradrenaline, or cortisol excretion rates (7). However, pronounced age-related changes in reaction to stress and level of fatigue were found in females in occupations involving monotonous and intensive work (16). Differences were found between age groups in the perceived degree of physical and mental load at work and in the subjective feeling of fatigue. Workers over 40 years of age most often perceived the mental and physical work load to be considerable. No significant correlation existed between age and cortisol and catecholamine excretion rates. Further large-scale field studies on age-related changes in responses to stress would be of considerable interest.

## Gender and stress at work

Data on gender-related differences in responses to stressors are relatively scarce, most having resulted from the work of one group of investigators.

Performance in different kinds of stress situation has been studied—in an intelligence test under pressure of time (19), in carrying out a difficult cognitive task (10), in an examination (12, 30), in carrying out a cognitive-conflict task (6), and in an achievement situation characterized by high controllability (11). Some of the more important results of these investigations are presented below.

145

There are no sex differences in cortisol and catecholamine excretion rates in persons at rest and under non-stressful working conditions. During psychological testing—an intelligence test under pressure of time and a difficult cognitive task— no significant differences in performance levels were found between men and women. In spite of the higher subjective arousal level in the women, adrenaline excretion increased only in the men. During examination stress a greater increase in the adrenaline and 4-hydroxy-3-methoxyphenylethylene glycol (MHPG) excretion rates was found in men than in women.

It was concluded that women cope with stress in a physiologically more economical way but at a higher psychological cost (12). The women complained more of discomfort and dissatisfaction with their achievements; the men expressed feelings of success and perceived effort more often. Similar characteristics were found in the patterns of neuroendocrine response to stressors during the performance of a cognitive-conflict task by male and female engineering students (6). Adrenaline and cortisol excretion rates were higher in the male students but no differences in the noradrenaline excretion rates were found. The magnitude of the differences was small. Recent data from a study of male adolescents confirm the sex differences in neuroendocrine and physiological responses to examination stress described above. The higher adrenaline excretion rates in the males was regarded as an indicator of a constructive effort to cope with a challenging situation rather than a concomitant of distress (30). Differences in the motivation patterns between the groups were also taken into consideration; "male" stress response could be found in an ambitious female.

Data on age- as well as gender-related differences in reactions to stress are limited. Studies on the catecholamine excretion rates in men and women of different ages, in conditions of experimentally induced stress, established an increase in urinary adrenaline in all groups except the young women (2). The urinary noradrenaline levels increased mainly in older men and women. The reduced response of the sympathetic-adrenal-medullary system in young women could be explained by differences in age.

It can thus be concluded, from studies in experimental and real-life stress situations, that there are clear-cut gender-related differences in neuroendocrine and psychological responses to stress, mainly in young men and women—adolescents and young adults.

Further investigations are needed to clarify the differences in responses to stress between different age groups under natural working conditions.

## References

1 ANDRES, R. & TOBIN, J. Endocrine systems. In: Finch, C. B. & Hayflick, L., ed. *The handbook of the biology of aging.* New York, Van Nostrand Reinhold, 1977, pp. 357–378.

2. ASLAN, S. ET AL. Stress and age effects on catecholamines in normal subjects. *Journal of psychosomatic research,* **25**: 33–41 (1981).

3 BARNES, R. ET AL. The effects of age on plasma catecholamine response to mental stress in man. *Journal of clinical endocrinology and metabolism,* **54**: 64–69 (1982).

4 BAR-OR, O. Age-related changes in exercise perception. In: Borg, G., ed. *Physical work effort.* Oxford and New York, Pergamon Press, 1976, pp. 255–266.

5 BLICHERT-TOFT, M. Secretion of corticotrophin and somatotrophin by the senescent adenohypophysis in man. *Acta endocrinologica supplementum,* **78**: Suppl. 195 (1975).

6 COLLINS, A. & FRANKENHAEUSER, M. Stress responses in male and female engineering students. *Journal of human stress,* **4**: 43–48 (1978).

7 DALEVA, M. & HADŽIOLOVA, I. [Biochemical aspects of the strain of teachers' work.] *Problemi na higienata,* **V**: 28–36 (1980) (in Bulgarian with a summary in English).

8 EVERITT, A. V. The neuroendocrine system and aging. *Gerontology,* **26**: 108–119 (1980).

9 FINCH, C. E. The regulation of physiological changes during mammalian aging. *Quarterly review of biology,* **51**: 49–83 (1976).

10 FRANKENHAEUSER, M. ET AL. Sex differences in sympathetic-adrenal medullary reactions induced by different stressors. *Psychopharmacology,* **47**: 1–5 (1976).

11 FRANKENHAEUSER, M. ET AL. Dissociation between sympathetic-adrenal and pituitary-adrenal responses to an achievement situation characterized by high controllability: comparison between Type A and Type B males and females. *Biological psychology,* **10**: 79–91 (1980).

12 FRANKENHAEUSER, M. ET AL. Sex differences in psychoneuroendocrine reactions to examination stress. *Psychosomatic medicine,* **40**: 334–343 (1978).

13 FROLKIS, V. [*Aging, neurohumoral mechanisms.*] Kiev, Izdatel'stvo "Naukova dumka", 1981 (in Russian).

14 GREGERMAN, R. & BIERMAN, E. Aging and hormones. In: Williams, R., ed. *Textbook of endocrinology,* 5th edition. Philadelphia, Saunders, 1974, pp. 1059–1070.

15 GUSSECK, D. J. Endocrine mechanisms and aging. *Advances in gerontological research,* **4**: 105–155 (1972).

16 HADŽIOLOVA, I. ET AL. Age-related changes in work strain and fatigue. In: *Abstracts from the proceedings of an international symposium on work physiology, Sofia, 4–6 October 1982.* Sofia, Institute of Hygiene and Occupational Health, 1983, p. 10.

17 HALBERG, F. & NELSON, W. Chronobiological optimization of aging. In: Samis, H. V. & Capobianco, S., ed. *Aging and biological rhythms.* New York, Plenum Press, 1978, pp. 5–56 (Advances in experimental medicine and biology, Vol. 108).

18 HERZOG, A. & RODGERS, W. The structure of subjective well-being in different age groups. *Journal of gerontology,* **36**: 472–479 (1981).

19 JOHANSSON, G. & POST, B. Catecholamine output of males and females over a one-year period. *Acta physiologica Scandinavica,* **92**: 557–565 (1974).

20 KAACK, B. ET AL. Changes in limbic, neuroendocrine and autonomic systems, adaptation, homeostasis during aging. In: Ordy, J. M. & Brizzee, K. R. ed. *Neurobiology of aging: an interdisciplinary life-span approach.* London, Plenum Press, 1975, pp. 209–231.

21 KÄRKI, N. The urinary excretion of noradrenaline and adrenaline in different age groups, its diurnal variation and the effect of muscular work on it. *Acta physiologica Scandinavica,* **39**: 7–96 (1956).

22 KEUTMANN, E. & MASON, W. Individual urinary 17-ketosteroids of healthy persons determined by gas chromatography: biological and clinical considerations. *Journal of clinical endocrinology and metabolism,* **27**: 406–420 (1967).

23 KOGAN, M. ET AL. [Excretion of glucocorticoids and their metabolites, 17-ketosteroids, aldosterone, electrolytes and catecholamines in practically healthy people in relation to age and sex.] *Laboratornoe delo,* **2**: 78–81 (1971) (in Russian).

24 MELTZER, M. The reduction of occupational stress among elderly lawyers: the creation of a functional niche. *International journal of aging and human development,* **13**: 209–219 (1981).

25 NAEGELE, G. Die Problematik älterer Arbeitsnehmer aus sozialpolitisher Sicht. [The problem of elderly employees from a sociopolitical point of view.] *Zeitschrift für gerontologie,* **8**: 438–452 (1975) (in German).

26 NAVAKATIKIAN, A. & KRIJANOVSKAJA, V. [*Age-related changes in metal working capacity.*] Kiev, Zdorov'e, 1979 (in Russian).

27 NELSON, W. ET AL. Rhythm-adjusted age effects in a concomitant study of twelve hormones in blood plasma of women. *Journal of gerontology,* **35**: 512–519 (1980).

28 PALMER, G. J. ET AL. Response of norepinephrine and blood pressure to stress increases with age. *Journal of gerontology,* **33**: 482–487 (1978).

29 PRINZ, P. ET AL. Circadian variation of plasma catecholamines in young and old men: relation to rapid eye movement and slow wave sleep. *Journal of clinical endocrinology and metabolism,* **49**: 300–301 (1979).

30 RAUSTE-VON WRIGHT, M. ET AL. Relationships between sex-related psychological characteristics during adolescence and catecholamine excretion during achievement stress. *Psychophysiology,* **18**: 362–370 (1981).

31 RENNER, V. ET AL. Stress and coping strategies over the life span. In: *Abstracts from the Proceedings of the 12th International Congress of Gerontology, Erlangen-Nürnberg, July 1981.*

32 SEVER, P. ET AL. Plasma-noradrenaline in essential hypertension. *The lancet,* **21**: 1078–1081 (1977).

33 SVECHNIKOVA, N. & BEKKER, V. In: [*Problems of neuroendocrine pathology and gerontology.*] Moscow, "Gorkij," (1967) (in Russian).

34 SVECHNIKOVA, N. & BEKKER, V. [Functional condition of the adrenal cortex in the process of aging.] *Problemy ēndokrinologii,* **16**: 3–7 (1970) (in Russian).

35 TIMIRAS, P. *Developmental physiology and aging.* New York, Macmillan, 1972.

36 TORSVALL, L. ET AL. Age, sleep and irregular work hours. *Scandinavian journal of work, environment and health,* **7**: 196–203 (1981).

37 UOZUMI, T. ET AL. Urinary steroids of healthy subjects from childhood to old age. *Acta endocrinologica,* **61**: 17–24 (1969).

38 ZIEGLER, M. ET AL. Plasma noradrenaline increases with age. *Nature,* **261**: 333–334 (1976).

# Part five

# Coping with stress and promoting health

# Chapter 15

# Health promoting factors at work: the sense of coherence

Aaron Antonovsky[1]

## The misleading pathogenic focus on stressors

The pathogenic approach that characterizes thinking, research, and action in the area of what is called "health" but which should, given the approach, be called "disease", is a major impediment in work for health (2, 3). At the centre of such an approach is a concentration on the outcome of disease and an overwhelming concern with pathogens. The same criticism applies in the field of occupational stress (24). It is imperative that a salutogenic approach should acompany the pathogenic approach in order to ensure that individuals and groups of individuals are able to cope successfully with pathogens and move towards, or at least not away from, a state of health within the health–disease–health continuum. When dealing with health promoting factors at work there are four major reasons for adopting this position.

Firstly, potential pathogens—in the present context, psychosocial stressors—at work are always ubiquitous. For example, while writing this chapter, my stressor load is considerable, even disregarding the stressors in my extra-working life: will the deadline be met; will the quality of the work be deemed satisfactory by editors, co-authors, and readers; do I really know what I want to say; is it worth saying; how will my other commitments be met if I spend time on this task? The telephone has to be answered; my new secretary needs guidance; meetings have to be attended; a junior colleague must discuss a bright idea immediately. The list could be extended considerably.

Few of the stressors mentioned above were regarded, in a work published in 1980, as being among the major classes of psychosocial stress facing workers (22). Compared with many others, I am a most fortunate worker; yet my stressor load is far from the minimum. No one's work stressor score, if measured seriously, would be low in an absolute sense. Regardless of this, efforts to reduce occupational stressors have not been insignificant, as is evident from the considerable body of material demonstrating their role in the etiology of physical and mental disease. Stressor loads do, however, differ. There are characteristics of a working situation that some would perceive as

[1] Department of the Sociology of Health, Centre for Health Sciences, Ben Gurion University, Beersheba, Israel.

stressors while others would not. It is thus unrealistic to think that a working environment totally free of stressors could be created; people will always have to cope.

Secondly, the pathogenic approach causes stressors to be viewed as bad for health in a theoretically and empirically indefensible way. An acceptable definition of a stressor is that it is a *demand that taxes or exceeds the resources of the system* or, to put it in a slightly different way, a demand to which there is no readily available or automatic adaptive response (*15*). Certainly some psychosocial stressors are so powerful that they must be regarded as noxious; but many can be overcome successfully, leaving no noxious effect; rather providing a positive experience. Many can give rise to a new adaptive response. Tension—the initial response of the body to a stressor—may well be salutogenic (*1, 2, 6*), illustrating once again the major significance of coping.

Thirdly, a pathogenic approach leads to a disregard of the salutogenic characteristics of a job. The question that tends to be asked in relation to health is what is bad about a job and not whether there is anything good about it. At first sight, it would seem that, from a human point of view, the first question is of most concern, but if the term "good" is conceptualized to mean the availability of resources for coping, it may well be that if the last question were to be of most concern the health consequences would be more effective. For example, it might well be more important, for an understanding of the health outcome, to know that a job is a union job than that the incumbent would be exposed to work overload. There is no word in the professional vocabulary, analogous to the word "stressors", describing health promoting factors. If a stressor leads, under certain conditions, to stress and disease, what leads, under certain conditions, to the management of tension and health? Perhaps the terminology used by Selye, who referred to "dis-stressors" and "eu-stressors", could be adopted (*23*).

Fourthly, a pathogenic approach encourages the piecemeal identification of individual stressors. The literature is replete with intuitively derived, only at times empirically supported, lists and classifications of occupational stressors. One author lists nine major classes of psychosocial stress facing workers; another divides the sources of occupational stress into three domains—environmental factors, job characteristics, and career considerations—each one followed by examples (*11, 22*). As other authors have said: "What seems to be lacking in this research is an integrating theoretical framework for stress-related job characteristics that can be assessed for the full workforce" (12).

Such a theoretical framework can best be formulated in terms of ways of coping with stressors. Such a theory, supported by empirical evidence, will enable the job characteristics that enhance the capacity to cope and those that weaken it to be determined, and well-directed action programmes to be developed. Without such a theory, the realm of bright ideas remains, with some fortunately, able to be applied pragmatically, but others causing resources to be devoured to no purpose.

When children worked 16 hours a day in dangerous mines, sophisticated theoretical models for coping with psychosocial stressors would have been superfluous. Unhappily, significant parts of the labour force, even in industrialized countries, still work under conditions patently destructive to health. There is much evidence that for a substantial segment of the labour force, even when efforts are made to ameliorate brutal and ugly conditions, both pathogenic and deleterious job characteristics remain. It is, therefore, imperative that a testable theoretical model should be developed, with a view to understanding and possibly exploiting the characteristics of life at work that promote health.

## The sense of coherence

Research activities over many years and an analysis of the literature on coping with psychosocial stressors have led to the development of the *sense of coherence* construct as the core of a model for coping (2, 3). The central hypothesis in the model is that the higher the location of the individual or group of individuals on the sense of coherence continuum, the more adequate will be the capacity to cope with the psychosocial stressors inevitably posed by the internal and external environments.

The sense of coherence is explained as a generalized emotional-cognitive perception on the part of an individual of the stimuli bombarding him, as they are, to a greater or lesser extent, controlled by him. The stimuli are seen as comprehensible, manageable, and meaningful; as information rather than as disquiet. If stressors are viewed as entropic, and inherent in human existence, then the sense of coherence, with its three inextricably interwoven components, represents the forces of negative entropy.[1]

*Comprehensibility* refers to the extent to which the stimuli confronting an individual are perceived by him as making cognitive sense; whether the information is ordered, consistent, structured, and clear—and hence predictable—or disquieting, chaotic, disordered, random, or accidental—and hence unpredictable. There is less reluctance to enter an open-ended situation, since there is confidence that sense and order can be made of it. A high degree of comprehensibility is necessary, but not sufficient in itself, for a strong sense of coherence; the stimuli may be clear but they pose demands.

*Manageability* refers to the extent to which the individual perceives that the resources at his disposal are adequate to meet the demands posed by the stimuli. This concept seems similar to that of control, widely used in the study of occupational stress, which derives from Rotter's locus of control theory (*20*). However, there is a crucial difference in that Rotter, and those whose work derives from his, think that resource control must be in the hands of the person seeking to cope. This is indeed one possibility. But it is also possible to achieve a high degree of manageability when resources for coping are legitimately seen as being controlled by other well-disposed and reliable persons or images—friends, workmates, a leader, or God.

When computers or organ systems are confronted with comprehensible stimuli and have the resources for managing them appropriately, the problem is potentially solved. But man is different; he must *care* to cope. *Meaningfulness* refers to the extent to which an individual feels that life makes sense, emotionally as well as cognitively; that at least some of the problems and demands encountered are worth an investment in energy, commitment, and engagement, and are welcome challenges rather than burdens.[2]

Although the plausibility of the hypothesis which, in the light of research, relates the sense of coherence to health status, is considered at length elsewhere in the literature (*2, 13*), no more

---

[1] Thinking of the sense of coherence in these terms immediately suggests an analogy to the physical, chemical, and biological processes. In its broadest sense, it addresses the problem central to all science: the transformation of disorder into order. The link to immunology and neuroendocrinology is clear. Eventually, the sense of coherence construct may not only be analogous to, but subsumed within, a comprehensive theory of negative entropy.

[2] A 29-item questionnaire designed to provide an operational measure of the three components of the sense of coherence is available from the author.

than preliminary data are at hand to test it directly (4, 21). The way the sense of coherence actually works to cope successfully with psychosocial stressors is also dealt with elsewhere (3). Suffice it to say that, though a strong sense of coherence is considered to be significant in (1) avoiding stressors, and (2) appraising stimuli that cannot be avoided as opportunities rather than stressors (16, 19), its major function is to cope directly with stressors, preventing initial tension from being transformed into stress.

## The development of the sense of coherence

The sense of coherence construct emerged from a search for a clear answer to the question: How does the possession of resistance resources eventuate in good health? What such resources,—e.g., money, a clear ego identity, cultural stability, social support—have in common is that they repeatedly provide life experiences that have three characteristics: consistency, underload-overload balance, and opportunities to participate in decision-making. From earliest infancy to young adulthood man undergoes countless experiences characterized, to a greater or lesser degree, by:

— surprises and contradictions that cannot be reasonably explained;
— violations of individual "critical tolerance" loads at either end of the load spectrum (17);
— little choice in the tasks that have to be undertaken, little performance responsibility, and little influence on the outcome.

The phrasing is negative; it can, of course, be positive. By young adulthood, these experiences, having been shaped by the family, subculture, and social structure in which the individual has grown up, have given rise to a tentative generalized way of looking at the world, which is called the sense of coherence. There are parallels between consistency and comprehensibility, underload-overload balance and manageability, and opportunities to participate in decision-making and meaningfulness.

It is suggested that the experiences of childhood and adolescence considerably influence the initial, tentative location in the sense of coherence continuum. While this location determines, in some measure, the manner of entering into this or that adult family situation, or this or that working situation, it is the experiences in the first decade or so of adult life that are decisive in finally determining the strength or weakness of an individual's sense of coherence. This suggestion, if it is correct (and if the sense of coherence–health hypothesis is similarly correct) immediately leads to a major conclusion with regard to health-promoting factors at work, namely, that priority attention should be given to the working conditions of the young. With a strong sense of coherence at the age of 30 years there is good reason to believe that the worker will be able to cope with the subsequent vicissitudes of life. Yet unemployment—the life situation most destructive of the sense of coherence—is widespread among the young.

Having stressed the fact that the sense of coherence is likely to rest at the same level after young adulthood, it should be noted that minor modifications can occur throughout life. Such modifications can make tangible differences to health fates. While undramatic, incremental change during the course of the life cycle cannot be disregarded, radical, major changes in working conditions may, even for older individuals, substantially change the strength of their sense of coherence. The middle-aged housewife who turns to university studies, takes up gratifying work, or goes through a divorce which ultimately proves to be liberating may slowly, and only after some years of experiences that contrast sharply with those of previous decades, emerge with a much strengthened sense of coherence. The 50-year old research chemist, who finds himself unemployed and isolated from active society, may move in the opposite direction. Migrant farm labourers, who become involved in a strong union which radically changes their conditions of life making them more stable, or those who set up autonomous worker-owned plants, may come to see the world differently.

There is no magic formula to work miracles overnight, but suffering can be assuaged and wellbeing increased if there is adequate understanding of the ways in which working conditions can shape the sense of coherence and, through it, the health status.

## Meaningfulness, manageability, and comprehensibility at work

The argument up to this point has been that a statutory approach and the sense of coherence concept provide a theoretical model for the analysis of working conditions, through which the sense of coherence, and thus the ability to cope with stressors, may be strengthened or weakened. This is contrary to the traditional, non-theoretical focus on occupational stressors or dis-stressors. Some scientists have adopted a theoretical approach using the *fundamental human needs* concept (*10*). Marxist students of work alienation and its relation to illness similarly start out from theory (*18*). Whichever model proves to be the most powerful, it is wise to start from theory rather than to continue to think in *ad hoc* terms.

By asking an individual worker, a group of workers, or members of an occupational group, a social class, or the labour force as a whole, how working conditions contribute to their feelings in regard to comprehensibility, manageability, and meaningfulness, a better understanding can be gained of the health consequences of work. No single study can provide a full test of the sense of coherence theory, since the sample is always limited, and the data incomplete. For example, a study published in 1981 was restricted to a sample of the Swedish male workforce and to the variables representing manageability and meaningfulness (job demands and job decision latitude) (*12*). Each individual study, however, forms part of the continuing research programme.

It is not necessary to be a Marxist to be aware that, almost without exception, the literature on occupational stress deals objectively with immediate job conditions and subjectively with the ways in which those conditions are perceived, with complete disregard for the historical and broader social structure within which the job is embedded. The sense of manageability could conceivably be understood by studying only the immediate work process but to understand comprehensibility in that context alone would be difficult, and to understand meaningfulness would be impossible. For example, the data from a study of occupational stress and hospitalization rates in United States Navy enlisted men over a 30-year period were analysed without mention of the fact that the period of the study included the Viet Nam war years (*11*). Was the sense of coherence for those men not influenced by whether the country was at war or at peace? A social-historical awareness, without which research will be seriously flawed, should lead to the types of questions described below.

*Meaningfulness.*

Participation in decision-making is essential to a feeling of meaningfulness at work. This can be expressed as *joy and pride in work* (*10*) or *discretionary freedom.*

If an individual has freely chosen the type of work he does, instead of having fallen into it, he will experience joy and pride in what he is doing, feeling a sense of ownership in the work and that he wishes to do it. What determines joy and pride in work? Intellectuals, writing about work, tend to see self-expression as the central issue. Without disregarding the importance of self-expression, social valuation at two levels is of greater significance for most workers. Firstly, there is the social valuation of the *enterprise*—the occupation, the industry, the factory. Teachers in the pre-Israel Jewish community of Palestine can be cited in relation to this level (5), as well as the situation of the housewife, who does not work in the accepted sense of the word. Such valuation is expressed in the resources allocated to the enterprise by the society in which the worker lives—power, rewards, prestige. Secondly, there is the social valuation given by the *individual* worker himself; the more he perceives that the social valuation of his work meets his own criteria of equity, the more is he likely to feel a sense of ownership for it. There are, of course, other contributions to the meaningfulness of work—the intrinsic gratifications, which are inextricably linked to manageability—as well as the sense that the rewards of work enable other values to be realized, such as, first and foremost, being able to support a loved family.

In the study of occupational stress, discretionary freedom is most often thought to refer to the decision latitude given to the individual worker. Indeed, the worker who feels that he is free to choose his tasks, their sequence, and at what pace he is to perform them, is likely to see the work as being meaningful. Having a voice in what tasks are to be performed leads to the wish to invest energy—the core of meaningfulness—in them. This, however, disregards the crucial collectivity of most work; of no less significance for meaningfulness is the extent to which the worker has a voice in what is happening around him. This leads to the question of how one individual's work relates to that of others; is their relationship detached or complementary, antagonistic or cooperative? An analysis of the meaningfulness, in the context of the alienation theory, of the work of nurses in two different social structures was published in 1964 (7). A further question to be asked is what voice does the worker have in the overall production process at local and societal levels? The word "voice" means the power of the worker to influence what he himself does and what he considers of consequence in what is happening around him. This does not necessarily imply a monopoly of control. The decisive issue is whether the worker perceives that control is legitimately lodged, so that what he regards as an appropriate contribution is being made to the decision-making as part of a collective.

## Manageability

An appropriate overload-underload balance is seen as being decisive in determining the sense of manageability. Some of the issues related to control have been discussed above; but control is also associated with manageability. The more problems an individual poses for himself, or the more posed by someone he recognizes as legitimately entitled to pose them, the more likely is he to feel that he has the resources to solve them. For the worker, overload relates to having or not having the resources at his disposal to deal with problems successfully, whoever poses them; that is, basically, the knowledge, skills, material, and equipment he needs. The argument must, however, be carried two steps further. Firstly, the formal social structure in which most work is embedded must be perceived to provide the appropriate environment and equipment needed to carry it out well. Virginia Woolf said that all a woman novelist needed to work well was a room of her own and £50. In most work settings, the feeling of working well comes from the perception that others on whom the work is dependent are also working well: those working on the product at every stage, those providing the material and equipment, and those organizing the work. Secondly, in most working situations informal social structures come into being. At times, the character of social relations is of even greater importance than the worker's own resources or those that the formal structure places at his disposal: if he is feeling unwell, can he count on others to take over; if he makes a mistake, can he count on others to understand and help to rectify it?

Perceived resources, therefore, provide the key to coping with the problem of overload. This discussion of manageability refers only to resources. The assumption is that the worker considers the problems confronting him to be legitimate and appropriate to his work; if he does not, the problems impinge upon meaningfulness and comprehensibility. The resources in question may also be collective or extra-individual. What impinges on the sense of manageability is chronic or frequently repeated acute overload. The worker has no opportunity to rest and recuperate. Overload remains the main factor in regard to manageability, particularly if the frequently ignored data on "moonlighting", shift work, and those, such as housewives, who are doing two jobs while only being paid for one, are kept in mind. However, soldiers in battle, surgeons operating, scientists in their research activities, or pioneers having to drain swamps, are able to call on untapped resources, enhancing their sense of manageability. The discussion of Engel & Schmale on the conservation-withdrawal hypothesis is most relevant to overload and underload (9).

Underload may be equally devastating to the sense of manageability. The knowledge gained in psychology regarding sensory deprivation has been too long ignored (17). The consequences of boredom, tedium, and monotony, which impinge on the sense of meaningfulness, are not the primary concern; within limits these are functional (9). The core of the matter is that experience at work, allowing abilities to be exercised and potentials to be fulfilled, promotes confidence in being able to manage. This is the most significant implication of the work of Kohn and his colleagues on substantive complexity (14). They showed the long-range implications, for a variety of personality dimensions and a sense of distress, of the extent to which a job is characterized by demands that require substantive complexity to be performed adequately. Though their discussion disregards the danger of overload due to job complexity, there is little question that, for all too many—particularly minority groups, women, and those labelled as handicapped—a lack of substantive complexity at work, disregarding the individual's potentials, leads to an increasing paralysis of the sense of manageability.

*Comprehensibility*

Things that fit together, unknowns satisfactorily explained, and ordered patterns strengthen the sense of comprehensibility. To quote a personal example: in 1972, the author joined a very small group to plan a medical school which was to be based on considerably different values, conceptions, and goals from those characterizing a traditional medical school. He took part in the selection of students and taught them for six years, passing on the knowledge that emerged from his own research and keeping himself informed of the other things they were learning. Today, he is still in contact with them as a formal consultant, friend, or simple wellwisher, as they work as physicians and as some return to the medical school as young teachers. This continuing involvement of the school faculty and its graduates in working towards a transformed health care system involves the author as an individual in the broader social structure. The entire working experience is coordinated in a way that makes sense to him cognitively and emotionally. In contrast, Charlie Chaplin, as the character he portrayed in the film "Modern Times", performed a single task on the production line without the slightest notion of what he was helping to produce.

Coser's empirical and theoretical work provides an understanding of how the social structure of an enterprise is decisive in shaping experiences that enhance comprehensibility (7, 8). In his empirical study, nurses working at a rehabilitation centre were compared with nurses in a custodial hospital ward. The nurses at the rehabilitation centre were in frequent structured contact—often in conflict—with other staff members, confronting common problems; they interacted with patients and their families; they were working towards long-range goals. The custodial nurses worked as individuals, focusing on physical order and cleanliness. In the theoretical study, the concept of structured role complexity was examined; this is the counterpart of Kohn's concept of substantive job complexity. Familiarity with the roles of others in the working context, with possible alternative solutions to problems, with overall goals, and with plans helps to provide the worker with a comprehensible picture of his working environment. Again, however, the danger of over-complexity, resulting in chaos, must be avoided.

Whatever the importance of structured complexity, it is less fundamental to comprehensibility than a working condition so obvious in its significance that it often tends to be disregarded, namely, job security. In relating his experience at the medical school, the author takes it for granted that it is understood he has academic tenure; his job, with its quite respectable rewards, is assured until he retires; it provides for a reasonable pension; and his skills are such that he can go on working for as long as he chooses; whatever the social developments, the university will continue to exist. Thus, not only are his present circumstances clear but the future is predictable; there will certainly be surprises, new problems—some, such as cuts in budget, will not be welcome—but chaos will not be the outcome.

The question of job security must be considered at four levels:

(1) The individual worker's calm belief that, as long as the rules accepted as legitimate are not violated, he will not be dismissed.

(2) Whatever his competence, his confidence that the type of work or the section in which he is employed will not come to be defined as redundant, without it having been anticipated and alternatives for retaining his services planned.

(3) For both employed and self-employed workers, the viability of the enterprise, i.e., whether it is profitable or not.

(4) The confidence he feels in the continuity of the social system, a concern that extends beyond the world of work but which, in an era of violent outbreaks and the threat of nuclear warfare, cannot be ignored.

Although each of these issues would need a full chapter for adequate discussion, no consideration of the working experiences that promote the sense of comprehensibility can be serious unless they are at least noted.

A further variable to be considered is the nature of the social relations within the working group. Shared values, a sense of group identification, and clear normative expectations, lead to an ambience of consistency. In theory, there need be no contradiction between an ambience of consistency and a hierarchical, bureaucratic, structuring of work. In practice, however, if role allocation is not based on functional performance and if functional authority spreads to irrelevant areas, solidarity is disrupted. Further, difficulties in communicating and in identifying within the working group are exacerbated if it is not culturally homogeneous and if, as is often the case, prejudice exists.

When, therefore, the environment at work enables the worker to see the entire spectrum and his place in it, fosters confidence and the feeling of security, and supports communicability in social relations, his view of the world as being comprehensible is strengthened considerably.

## Concluding remarks

The overwhelmingly disproportionate emphasis in applied work is unwittingly illustrated in a paper on health promotion at the worksite, in which only part of one paragraph is devoted to procedures for helping people to change their working environment in order to improve their health, whereas nearly four pages are devoted to procedures for helping people to cope with an environment that cannot be changed (22). This is not a criticism but an illustration of the current situation, with all its ideological implications.

The entire purport of this chapter is to explain how the success of coping with stressors is primarily determined by the strength of both the individual and the group sense of coherence. True the strength of the sense of coherence is largely shaped by life experience, particularly for older workers, but it can be modified, detrimentally or beneficially, by the nature of the current working environment. The discussion has, therefore, related primarily to the theoretical model, which enables transformations in the working environment to be consciously designed with the purpose of enhancing the workers' sense of comprehensibility and meaningfulness.

The Utopian notion of a stressor-free environment is rejected as impossible. However, Frankenhaeuser's use of the concept of *healthy maladjustment* to describe the failure of man to adapt to inhuman conditions can be supported. Such pseudo-adaptation permits subtle symptoms of maladjustment to accumulate gradually without eliciting corrective responses on the political and cultural levels (*10*). The focus must thus be on the working environment and not on procedures for managing stress.

A distinction must be made between the elimination of stressors and the development of health-enhancing job characteristics. The former is certainly of importance and to some extent possible; the latter, which stems from a salutogenic approach, is of crucial importance. The distinction between the two is far from absolute. There are job characteristics that promote the sense of coherence the absence of which could reasonably be termed a stressor—e.g., a harmonious, complementary pattern of a working relationship between a group of doctors and nurses, as opposed to a relationship full of conflict. There are also job characteristics that promote the sense of coherence but the absence of which would generally not be termed a stressor. Thus, if the nurses and doctors working in a group are alienated from each other, or if a factory worker cannot see his role in the overall production process, tension and stress would not necessarily be the direct response. A pathogenic approach, without a theoretical base, would place emphasis on the elimination of stressors; the theoretical model discussed in this chapter places emphasis on the promotion of health.

No doubt the sense of coherence construct requires refinement and testing but it is an approach consistent not only with the growing consensus of opinion regarding the importance of coping but with the theoretical ideas advanced in the recent work of eminent researchers (*10, 13, 19*). This type of approach is also being discussed in association with knowledge of the central nervous and other systems of the body, the pathways linking stressors to health.

# References

1 ANTONOVSKY, A. Social and cultural factors in coronary disease. *Israel journal of medical sciences,* **7**: 1578–1583 (1971).

2 ANTONOVSKY, A. *Health, stress and coping: new perspectives on mental and physical well-being.* San Francisco, CA, Jossey-Bass, 1979, pp. 92–97.

3 ANTONOVSKY, A. The sense of coherence as a determinant of health. In: Matarazzo, J. D. & Miller, N. *Behavioral health: a handbook of health enhancement and disease prevention.* New York, Wiley, 1984, pp. 194–202.

4 ANTONOVSKY, A. *The sense of coherence: development of a research instrument* (unpublished).

5 BEN-DAVID, J. Professionals and unions in Israel. In: Freidson, E. & Lorber, J., ed. *Medical men and their work.* Chicago, IL, Aldine, 1972, pp. 20–38.

6 COSER, L. A. *The functions of social conflict.* New York, The Free Press, 1956.

7 COSER, R. L. Alienation and the social structure: case analysis of a hospital. In: Freidson, E., ed. *The hospital in modern society.* New York, The Free Press, 1963, pp. 231–265.

8 COSER, R. L. The complexity of roles as a seedbed of individual autonomy. In: Coser, L. A., ed. *The idea of social structure: papers in honor of Robert K. Merton.* New York, Harcourt Brace Jovanovich, 1975, pp. 237–263.

9 ENGEL, G. L. & SCHMALE, A. H. Conservation-withdrawal: a primary regulatory process for organismic homeostasis. In: Porter, R. & Knight, J., ed. *Physiology, emotion, and psychosomatic illness.* Amsterdam, Associated Scientific Publishers, 1972, pp. 57–85 (Ciba Foundation Symposium 8).

10 FRANKENHAEUSER, M. Coping with stress at work. *International journal of health services,* **11**: 491–510 (1981).

11 HOIBERG, A. Occupational stress and illness incidence. *Journal of occupational medicine,* **24**: 445–451 (1982).

12 KARASEK, R. ET AL. Job decision latitude, job demands, and cardiovascular disease: a prospective study of Swedish men. *American journal of public health,* **71**: 694–705 (1981).

13 KOBASA, S. C. ET AL. Personality and constitution as mediators in the stress–illness relationship. *Journal of health and social behavior,* **22**: 368–378 (1981).

14 KOHN, M. L. & SCHOOLER, C. Job conditions and personality: a longitudinal assessment of their reciprocal effects. *American journal of sociology,* **87**: 1257–1286 (1982).

15 LAZARUS, R. S. & COHEN, J. B. Environmental stress. In: Altman, I. & Wohlwill, J. F., ed. *Human behavior and environment: advances in theory and research.* New York, Plenum Press, 1977, Vol. 2, pp. 89–127.

16 LAZARUS, R. S., & LAUNIER, R. Stress-related transactions between person and environment. In: Pervin, L. A. & Lewis, M., ed. *Perspectives in interactional psychology.* New York, Plenum Press, 1978, pp. 287–327.

17 LIPOWSKI, Z. J. Sensory and information inputs overload: behavioral effects. *Comprehensive psychiatry,* **16**: 199–221 (1975).

18 MANDERSCHEID, R. W. Stress and coping: a biopsychosocial perspection on alienation. In: Geyer, F. R. & Schweitzer, D., ed. *Alienation: problems of meaning, theory, and method.* London, Routledge & Kegan Paul, 1981, pp. 177–191.

19 PEARLIN, L. I. & SCHOOLER, C. The structure of coping. *Journal of health and social behavior,* **19**: 2–21 (1978).

20 ROTTER, J. B. Generalized expectancies for internal versus external control of reinforcement. *Psychology monographs,* **80**: 1–28 (1966).

21 RUMBAUT, R. G. ET AL. *Stress, health and the sense of coherence.* San Diego, University of California (unpublished).

22 SCHWARTZ, G. E. Stress management in occupational settings. *Public health reports,* **95**: 99–101 (1980).

23 SELYE, H. Confusion and controversy in the stress field. *Journal of human stress,* **1**: 37–44 (1975).

24 SHARIT, J. & SALVENDY, G. Occupational stress: review and reappraisal. *Human factors,* **24**: 129–162 (1982).

# Chapter 16

# Fitting work to human capacities and needs: improvements in the content and organization of work

Lennart Levi[1]

## Introduction

Until recently, the work-related health problems reviewed in the previous chapters have been managed by providing financial compensation for the occupational health risk and the resulting disease, and by intervening with surgical, pharmacological, and/or psychological measures. A gastrectomy to treat peptic ulcer in a worker exposed to a very high work-load in a highly repetitive task at an assembly line is an example of what intervention could amount to in practice; or it could take the form of supplying benzodiazepines to modify his emotional reactions or an anticholinergic drug to modify the flow of nervous stimuli to the acid secreting parts of his stomach; or he could be made to believe that his working situation is acceptable, or even the optimum, governed by natural laws, not accessible to change, and thus to be accepted and even enjoyed. Indeed, attempts such as this to make the "foot fit the shoe" cannot be dispensed with entirely. Some environmental conditions are difficult to change, at least for the time being, and the individual may need relief from his symptoms at once (*10, 11*). A necessary complementary approach to facilitating good person–environment fit, however, aims at adapting the "shoe to fit the foot" (*6, 9, 13*) and providing a variety of "shoes"—i.e., ecological niches.

An invalid in a wheelchair may lead a very isolated and depressing life, cut off from work and human contacts because he lives in an apartment without an elevator. The "modern" approach to managing his depressive reactions might be to administer antidepressant drugs and possibly to submit him to psychotherapy. Such *structural* obstacles to good health and a good quality of life are abundant, as are the pharmacological and/or psychotherapeutic "solutions". A better solution would be to help him find a type of housing that allows him to leave it in his wheelchair and go to work without having to be carried. This example illustrates that environmental approaches are needed, focused on prevention.

In order to improve person–environment fit at work, the preventive strategy in occupational medicine has generally made use of one of the three following approaches:

[1] WHO Psychosocial Centre, Laboratory for Clinical Stress Research, Karolinska Institute. Stockholm, Sweden.

(1) Adaptation of the work and the working environment to the worker's abilities, needs, and expectations;

(2) Elimination of, or protection from, a noxious working environment;

(3) Modification of the worker's psychobiological programme—i.e., his propensity to react in a certain manner—so that the working environment can be accepted, or at least tolerated, without provoking too much stress or causing illness (10, 11, 12). Whatever the approach, the individual, the group, and the environment must be regarded as components of a system in which each is affected by the others in many ways. Although this concept is one that appears to be self-evident, it is not applied today; it requires a systems approach in environmental protection, social planning, and individual care.

Some of the key questions to be considered in this chapter relate to whether it is possible, for instance, to improve the objective working environment or to alter the worker's experience and appreciation of that environment; whether the psychobiological programming can be influenced to take a favourable direction, so that the propensity to react pathogenically declines. Intervention to alter the mechanism, using drugs, has already been mentioned; perhaps it could be supplemented by psychotherapy or counselling, thus helping the worker to resolve his own conflicts and cope with his own problems, giving him a chance to talk things over with somebody who has time to listen, and offering him an opportunity to adapt his situation to his own abilities, needs, and expectations. Further questions relate to whether it would be worth while to try to prevent various diseases from developing by early diagnosis and prompt treatment; whether improvements in the environment outside working life serve to strengthen resistance to, or provide compensation for, the strains at work that cannot be avoided; and whether it is possible to protect specific vulnerabilities by endeavouring to put the right person in the right place.

## Principles for improving the working environment

### Basic approaches

Efforts to improve the working environment and to protect and promote the health and wellbeing of the worker should be based on:

(a) A *comprehensive* view of man and his environment—i.e., the equal and integral consideration of physical, mental, social, and economic aspects;

(b) An *ecological approach*—i.e., consideration of the interaction between the entire individual and the entire environment, physical, chemical, and psychosocial, and of the dynamics of the complex system;

(c) A *cybernetic* approach—i.e., the continuous monitoring and evaluation of the physical, mental, social, and economic effects on the worker of the working environment and of changes relating to it, so that it is continuously adapted and reshaped;

(d) A *democratic* approach—i.e., giving the worker the greatest possible influence over his own situation, as well as direct, efficient channels of communication to the decision-makers.

This, however, is not how most organizations function today. The application of occupational health knowledge to practice has not generally gone farther than attempts to optimize such parameters as the heights of tables and chairs, the lengths of reaching-distances, or the placement of knobs and dials, and existing ergonomical knowledge is not always applied even to those elementary questions.

The *comprehensive* approach entails consideration of the worker and the working situation not only medically, psychosocially, technologically, or economically, but from all those aspects. Consideration from a more restricted perspective might lead to unwise decision-making, as when a shift work system is designed exclusively with a view to facilitating the worker's digestion and sleep, disregarding the fact that he also has to function in the social context—interacting with family members, friends, and the community—or disregarding all those aspects in favour of productivity and profits.

The *ecological* approach means that full consideration is given to all components of the person—environment ecosystem. To neglect to do this can result in a decision to achieve full employment at any price, even if it leads to severe pollution problems at the plant and in the neighbourhood or, conversely, an absolutely pollution-free environment at the price of severe unemployment.

The *cybernetic* approach means that all the processes and the action taken are monitored and evaluated interdisciplinarily—e.g., following automation or the introduction of new legislation. This involves learning from experience and making the necessary adjustments.

The *democratic*, or participatory, approach allows individual workers and groups of workers to take part in making and implementing decisions. This has two major advantages. One is that, within reasonable limits, environmental adjustments can be influenced by the personal needs and preferences of the individual worker, increasing the chances of harmony through the resulting adjustments. The other is that the worker may see the actual process of being able to influence his own situation as something positive in itself and may, as a result, become more ready to accept the resulting environmental or individual adjustments.

## The "grassroot" approach

To adopt a "grassroot" approach means that much of the responsibility for reducing stress rests with the individual worker (5). By increasing the control he has over his job he himself can modify the demands of the job to bring it more into harmony with his individual preferences. This approach should not be confused with job enlargement or job enrichment. Proponents of the concept of job enlargement assume that every individual wants to be involved in his work and to have challenging experiences. The evidence supporting the person–environment fit theory suggests that to enlarge an entire set of job routines may improve fit for some, but it will worsen it for others, who prefer simpler routines. Increased control allows the worker to structure his job so that it better fits his preferences, whatever they may be. In an ideal situation, those who want more complex and challenging jobs can take advantage of the opportunities afforded, and those who prefer simpler jobs can choose to leave the decision-making to the others who want it. Accordingly, the "expert" and the "grassroot" approach could, and should, be combined. Different countries have chosen to combine them in different proportions.

## Types of organizational approach

In Europe, different countries have adopted various organizational approaches, conferring very different amounts of discretionary authority on the workers. In some, employees have the right to influence decisions on such factors as organizational and technological design, equipment, working methods, the types of material and products to be used, and personnel policies. In others, they also have the right to stop work if they think it is dangerous, or to seek expert advice on conditions they consider to be potentially hazardous (14).

The organization of worker resources is necessary but not sufficient. It is also very important, though perhaps difficult, to try to find ways to stimulate individuals to take care of their own problems. Narrow and system-paced jobs, in which the worker has little influence on management decisions, may foster a passive and alienated attitude that mitigates against him protesting even if he is exposed to a serious occupational health hazard. In this case the worker's awareness must be increased and his competence and power extended so that he is able to recognize and prompt changes in unhealthy working conditions. This is the aim of recent legislation in Scandinavia.

## The Norwegian Work Environment Act

The Norwegian Work Environment Act sets forth the following provisions:

(1) *General requirements.* Technology, work organization, work time (shift plans), and payment systems are to be designed in such a way as to avoid negative physiological or psychological effects for employees, as well as negative influences on the alertness necessary for the observance of safety considerations. Employees are to be given opportunities for personal development and for the maintenance and improvement of skills.

(2) *Design of jobs.* Jobs are to be designed in such a way as to include opportunities for employee self-determination and the maintenance of skills. They should include task variation and contact with others; monotonous, repetitive work, and work that is paced by machine or an assembly line, leaving no room for variation in work rhythm, should be avoided. There should be an understanding of the interdependence between the elements that constitute a job, and employees should be provided with information and feedback on production requirements and results.

(3) *Systems for planning and control.* Employees, or their elected representatives, are to be kept informed about the systems used for planning and control, such as automatic data processing systems, and any changes in such systems. They are to be given the training necessary to understand the systems and the right to influence their design.

(4) *Mode of remuneration and risk to safety.* Piece-rate payment and related forms of compensation are not to be used if salaried systems can increase the safety level.

### The Swedish legislative approach

The Swedish legislative approach is twofold. The 1977 Work Environment Law came into effect on 1 July 1978 (*18*). It contains general statements such as "Working conditions shall be adapted to man's mental and physical capacities" and "Jobs shall be designed so that the employees themselves may influence their working situation." These general principles are complemented by specifications from the National Board of Occupational Safety and Health and the provisions of the Codetermination Law (*17, 20*). A requirement under the Codetermination Law is that information on working conditions should be given to employee union representatives on all matters and at all levels. It entitles local unions to negotiate on any matter that may influence a job situation. The parties themselves—the managers and the employees at the local plants—must agree on suitable job specifications. To provide guidance, the Swedish Confederation of Trade Unions has endorsed a special action programme on the psychosocial aspects of the working environment (*21*).

## Efficacity of the approaches

Whether the laws described above and related recommendations will produce the desired results remains to be seen. They reflect many laudable sentiments but are not very specific. The general intent is expressed but what constitutes an offence is not specified. Much will depend on how the laws are applied and what are the costs involved. Whether the results of laws promulgated in the Scandinavian countries would be relevant to other nations is a question to be answered. It seems possible that, although countries differ considerably politically and in the way work is organized, lessons may be drawn from this compelling national experiment.

In summary, noxious agents in the work setting, be they physical or psychosocial, ought to be eliminated, by the management, by the unions concerned, or by the individual workers. When this cannot be done, vulnerable groups must be protected. These, and other goals described above, might be regarded as clearly utopian; no doubt they are in many parts of the world. Nevertheless, although they certainly cannot be reached overnight, they give direction to continuing endeavours.

Most managers would agree that the approaches described are economically sound, because they are likely to increase motivation and decrease absenteeism, staff turnover, and unrest. However, their adoption must never be solely for economic reasons but also, and primarily, for reasons of health, wellbeing, and democracy.

## Initiation of preventive measures

The Working Group for Mental Health Protection and Promotion of the Swedish National Board of Health and Social Welfare, outlined seven levels at which preventive measures should be initiated (19):

(a) The *structural macro* level—improving forms of work, introducing new methods of collaboration and codetermination, changing the ways in which public institutions function;

(b) The *structural micro* level—introducing a stimulating and safe environment in certain factories, offices, or other establishments;

(c) The level at which the individual worker's power of *resistance* can be strengthened—health promotion activities and training in the resolving of conflict and how to cope with problems;

(d) The level at which the individual worker's *adaptation to reality* can be improved—encouraging him to have realistic expectations with regard to working colleagues, job, and society as a whole;

(e) The level at which the concept of the *right person in the right place* can be promoted in a pluralistic society—vocational guidance;

(f) The level of crisis intervention and the *buffering of social support* during critical periods, especially for high-risk groups;

(g) The level at which the individual worker's *competence and power* to cope with his own and his neighbours' problems can be increased.

It is most important, in plans for the promotion of occupational health and the prevention of disease, for integrated measures to be introduced at each of the seven levels. To decide which measures should be taken is the task of government authorities. However, both individual workers and groups could, and indeed should, make important and decisive contributions.

## Guiding principles

The following guiding principles should be considered in efforts to prevent stress-related illness and to promote health and wellbeing, at work and elsewhere (10, 11, 12).

## Goals and choices

Of fundamental importance is the critical yet commonly neglected question of what are the *principal goals* of working life. The inclination has long been to stress quality; the technical and economic aspects of a job at the cost of the human aspects.

Health is defined as not merely the absence of disease or infirmity but also a state of complete physical, mental, and social wellbeing (*24*). The attainment of health in this broad sense must be one of the principal aims of all social activity, including working life and its conditions. Admittedly, working life also gives rise to income, the production of goods, and the provision of services, but they are not ends in themselves, only means of ensuring optimum physical, mental, and social wellbeing and promoting health, development, and self-realization. Thus, working life can satisfy human needs directly, through opportunities for creative and stimulating activities and social contacts, and indirectly, as a source of income (*11*).

There is a choice: either to allow the vast bulk of jobs that have to be done in the manufacturing and service industries to remain dull and monotonous as they are at present, accepting that work is a necessary evil to be endured—the principal aim then becomes a reduction in the amount that has to be done, with a shortening of working hours and the working week, while maintaining a pay scale that enables satisfaction to be sought elsewhere (*22*)—or to redesign the jobs and the organizational structures so that the majority, rather than the privileged few, can do meaningful and fulfilling work while maintaining a high level of performance (*15*).

## Psychological requirements

A redesign of job and organizational structures implies that the psychological requirements of the worker are met, other than those related to entitlements specified in the contract of employment, such as wages, hours of work, safety, or security of tenure. Six psychological requirements pertaining to the content of a job have been defined, which must be met if a new work ethic is to develop (*4*):

(1) The job must be found to be reasonably *demanding* in terms other than sheer endurance, and to provide at least a minimum of variety;

(2) It must be found possible to *learn* on the job, and go on learning;

(3) Responsibility must be felt in an area of *decision-making*;

(4) A degree of *social support* and recognition at the work place must be perceived;

(5) It must be possible to relate what is done and produced at work to *social life*;

(6) It must be felt that the job leads to a desirable *future*.

## Organizational consequences

Translated into organizational terms, measures to satisfy the psychological requirements defined above appear to involve two fundamental ideas: (1) A reversal of the trend towards the division of jobs into smaller and smaller elements, each to be performed by a single worker, and its replacement by a trend towards amalgamating various functions as a meaningful, integrated whole; (2) Modification of the hierarchical organizational structure by reorganizing workers into small face-to-face groups with a good deal of autonomy, with the supervisor no longer giving detailed orders but responsible for seeing that each group has the resources it needs and for managing its relations with the rest of the enterprise (23). Or, under the principle of self-regulation, only the critical interventions, desired outcomes, and organizational maintenance requirements need be specified by the managers, leaving the rest to the workers (22).

Neither of the above-mentioned measures, nor any others, provides an all-encompassing solution to the problems; in fact, there is no all-encompassing solution. What is good for one individual may not be good for another. What has to be done is to ascertain what is good—or what is bad—for whom, in what way, when, and under what conditions, after which it can be expected that interest will focus on more or less generally effective measures.

## Satisfaction of human needs

The tactics to improve person–environment fit at work concern the adaptation of environmental demands and opportunities to workers' abilities and needs.

The satisfaction of physiological needs, including the need to be secure and safe is often referred to as hygiene. Attention to hygiene, though extremely important, is not enough. There are other environmental factors connected with the satisfaction of other human needs. Such factors include salary, the number and quality of human contacts, supervision, security at work, and physical factors in the working environment (1). Unless the needs in connection with these factors are fulfilled to some extent, dissatisfaction and unhappiness will prevail. Once they have been satisfied, additional "improvements" will not substantially increase satisfaction.

Another class of factors is that referred to as motivational— i.e., related to needs such as ego-experience, self-esteem, and self-appreciation. Such factors include advancement to more stimulating tasks, appreciation of work well done, freedom to complete a task, authority to take responsibility, and the inherent qualities of the task itself. These factors have a positive influence on the worker's involvement in, and experience of satisfaction from, the work.

To be successful, therefore, an occupational mental health protection and promotion programme must be comprehensive and take into account all types of human need and the means of satisfying them through environmental and other adjustments— i.e., by creating a good person–environment fit. For such a fit to be good, the general organization of the work process is no less important than hygiene factors such as noise and illumination. Organizational factors are those primarily concerned with (3): (1) The worker's knowledge of how his part of the work contributes to the finished whole; (2) The independence and responsibility of the worker; and (3) His social contact and collaboration with other employees.

## The meaningful whole

The task should be seen to constitute a meaningful whole, or at least to form an essential part of a production process that is understandable and has meaning for the worker in terms of the relationship between his own contribution and the ultimate goals of the production process. Understanding could be achieved by allowing the worker to take part in the planning, realization, and control of the task. Another possibility might be to amalgamate fragmented pieces of the work process into a meaningful sequence.

The worker should be allowed to utilize his existing knowledge and skills through the work process and to develop them further also through the work process and through education, thereby enabling him to take over increasingly skilled tasks. This can be facilitated by choosing the right man for the right job and by rotating workers who wish to be rotated among different tasks and positions. Job rotation or enlargement can also involve increased participation for the worker in the planning and control of his own achievements. They should further include options—not necessarily demands—for recurrent training and education.

One of the most important ways of learning from experience is through feedback from the environment. Every worker ought to be aware of the results of his working endeavours—i.e., the quality and number of his achievements. This is true not only for those working individually but also for those working in groups. This can be achieved in several complementary ways. One is to allow continuous contact with those responsible for the preceding and subsequent stages in the production chain. Another is to allow the worker to control and check the results of his own work. In addition, his supervisors might regularly evaluate his performance quantitatively and qualitatively and prepare a report on it. There should also be opportunities for discussion with fellow workers and the managers of the organization on how the work is progressing (3).

## Independence and autonomy

For a worker to have independence and autonomy does not mean that he can do whatever he likes, whenever he likes, in any way he likes. This would be one extreme; the other would be to turn him into a passive tool, which is equally unfavourable. Work is much more stimulating, rewarding, and effective if an optimum degree of participation is allowed; for example in planning, or in deciding on working methods and pace, or the location of pauses. This can be achieved by making the individual worker, or the group, responsible for accomplishing the job, for when to have a break, and for deciding when a temporary absence from work is acceptable. Information on how the work is progressing will then allow each unit in the production chain to decide on how to proceed. This might mean that the pace of work has to be left to the individual or the group to decide. Responsibility and power should be delegated as much as possible to those directly concerned with the production process.

## Social contact and collaboration with other employees

One of the great advantages of working life is that it satisfies a basic human need by creating the social context for contact and collaboration with others. Accordingly, opportunities for conversation and contact between employees should be created as necessary parts of the production process, not eliminated. Whenever possible in planning, the various components of the work should be allocated to relatively small groups of, say, 4–8 people. Speech and eye contact, as well as collaboration, between the group members should be encouraged. If this is not possible for practical reasons the workers should be helped to meet each other during breaks, and opportunities afforded for friendly contact with supervisors and managers.

### Needs satisfaction as an integral part of the production process

In the ways described above, the satisfaction of human needs becomes an integral part of the production process. Working life is humanized and simultaneously becomes more smooth and efficient, with mutual solidarity and social support. Extreme, competitive individualism is counteracted in favour of a communal approach. Support from others may be tangible—i.e., giving concrete assistance in completing a task, or sharing resources. Or it may be psychological-emotional—i.e., lending a sympathetic ear, giving reassurance, or demonstrating concern and care (16). For these and other reasons the mental health of the executive is of particular importance, not only in itself but because of the effect it may have on the workers under him. For example, a mildly depressed manager cannot generate enthusiasm in his staff, and an irritable supervisor might be unable to provide support, instead creating stress for those under him. The most important environmental factor for man is his fellow man.

## Person–environment fit

### Psychosocial job characteristics

The 16 principal psychosocial characteristics of a working situation, which should be considered in an occupational environment and health programme, complementing the lists of physical characteristics found in most ergonomics textbooks, are described below. They can be measured by the corresponding scales that are included in the self-report questionnaire of the Michigan Organizational Assessment Package; the headings of the scales provide a reasonably comprehensive checklist (7):

(1) *Challenge*: the extent to which the worker's skills and abilities are involved.

(2) *Meaningfulness*: the importance and meaning of the task to the worker.

(3) *Responsibility*: the importance to the worker of performing successfully.

(4) *Variety and skill*: the number of tasks that make up the job and their complexity.

(5) *Task identity*: the worker's job in relation to the entirety of the product or service.

(6) *Task feedback*: the worker's knowledge, from the work itself, of how he is performing.

(7) *Work influence*: the worker's influence over decisions affecting his work.

(8) *Autonomy*: the worker's freedom to decide what shall be done to accomplish the job and how it shall be done.

(9) *Pace control*: the worker's control over the speed at which he works.

(10) *Role conflict*: incompatible demands in performing the job.

(11) *Role clarity*: the worker's knowledge of what is expected of him.

(12) *Task uncertainty*: the degree to which events and the procedures for managing them are predictable.

(13) *Task interdependence*: the extent to which the job requires coordination with others.

(14) *Role overload*: performance requirements in relation to time constraints.

(15) *Resource adequacy*: the availability of tools, supplies, and information.

(16) *Skill adequacy*: job demands in relation to the skills and training of the worker.

## Workers' differing abilities and needs

It is often taken for granted that equal human value means equal ability and needs. This is not so, since people differ in all possible respects. Fortunately, the range of differences is matched by the corresponding range of opportunities and demands to be encountered in working life, in the form of innumerable environments, requirements, possibilities, occupations, and careers. An important task, therefore, is to match the components of the ecosystem—i.e., to create opportunities for the right person to find the right job. With the aid of aptitude tests, job analyses, and vocational guidance, every person should be allowed to find an optimum, or at least acceptable, personal "ecological niche". When no such niche is available the alternative is to create it. Only as a last resort should the worker's expectations have to be adapted to the unfortunate reality.

Inevitably, this type of problem confronts the physically, mentally, or socially handicapped in particular (11). In a certain way everyone is handicapped to some degree and in some respect. For example, the majority of a randomly selected population will lack the psychological equipment necessary to become a cabinet minister, a university professor, or the manager of a large enterprise. Similarly, a great proportion would not be able to enter a career as an opera singer, a stevedore, or a fighter pilot. The question posed must not, therefore, be whether a person is fit or unfit but rather whether he is fit or unfit for a specific task under specified conditions (2).

Nevertheless, some generalizations can be made. A person with low back pain should generally avoid on occupation that entails carrying heavy burdens or working in an inconvenient position. A person with a propensity to gastric distress, sleep disturbance, or a nervous complaint might wish to avoid rotating shift work. It should be remembered that every individual has a breaking point which he should not be forced or induced to approach too closely. With increasing age, responsibility and work-load often tend to increase. Many elderly workers try to cope by mustering more and more of the resources available to them, making the safety margin smaller and smaller. Ideally, a worker should not be regarded as able or not able to perform the tasks assigned to him but as able or not able to perform them without paying an unreasonable price in terms of health and wellbeing.

When the way in which a worker has adapted to a job is evaluated there is a tendency to view his performance entirely from the vantage of the employer—i.e., in terms of his performance capacity. It is equally important to view it from his own point of view. The important and interesting question is not what a worker is unfit for, but how his abilities can best be utilized for his own, his employer's, and the community's mutual benefit.

It cannot be expected that a perfect person–environment fit will always be achieved; a certain amount of adaptation to environmental demands must be accepted. It could even be argued that too perfect a fit would deprive the worker of the challenges and difficulties that, after all, constitute an essential element of existence, and some unpreventable discrepancies might be compensated for by advantages in other areas. Yet it seems that currently too many workers are assigned to jobs for which they are not suited. There is even a tendency towards rationalization by eliminating the only ecological niches and working environments suited to some types of people; and those that remain are becoming more and more uniform for technical or economic reasons. Workers, though at risk of becoming deformed, are expected to accept and adapt, sometimes with the offer of economic compensation. There is clearly reason to question strongly the wisdom of such a system and such trends.

## Responsibility for the working environment

It is often assumed that responsibility for the humanization of working life lies only with selected professionals. It would be more logical and effective, however, for everyone to be involved. Some specific responsibilities would naturally have to be allocated since it is in the interest of all concerned to create and maintain an optimum working environment which will enable person–environment fit to be improved and, consequently, health, wellbeing, and productivity (11). Much can be accomplished through two universally available measurement and intervention tools, namely, listening and speaking. People know their own problems best and they should be encouraged to speak for themselves (8). If possible they should also be encouraged to adjust their immediate environments themselves, if they can do so without harming others.

## Evaluation

Some of the principles and measures reviewed in this chapter are well supported by current knowledge and could be translated into action without delay. If, at the same time, however, the workers are to be protected and the efficiency and effectiveness of the measures taken increased, they must be continuously evaluated, in order to allow lessons to be learnt from experience.

## References

1 BANERYD, K. Psykosociala effekter av fysisk arbetsmiljö och skyddsförha Illanden i arbetet. [Psychosocial effects of physical work environment and occupational safety.] Stockholm, Ministry of Labour, 1976, pp. 7–25 (document SOU 3) (in Swedish).

2 BOLINDER, E. Arbetsanpassning: praktisk information för skyddsombud m fl. [Work adaptation: practical information for labour protection officials and others.] Stockholm, Prisma Bokförlaget, 1974 (Landsorganisationen Informerar 4) (in Swedish).

3 BRÄNNSTRÖM, J. ET AL. Generella arbetsmiljökrav för Stalverk 80. [General demands on the work environment of Steel Plant 80.] Stockholm, Work Environment Laboratory, 1975 (in Swedish).

4 EMERY, F. E. Some hypotheses about the ways in which tasks may be more effectively put together to make jobs. London, Tavistock Institute, 1963 (Document No. T813).

5 HARRISON, R. V. Job demands and worker health: person–environment misfit. Dissertation abstracts international, 37: 1035B (1976). (Doctoral dissertation, University of Michigan (University microfilms No. 76–19, 150)).

6 KAGAN, A. R. & LEVI, L. Health and environment—psychosocial stimuli: a review. In: Levi, L., ed. Society, stress and disease: childhood and adolescence. London, New York, and Toronto, Oxford University Press, 1975. Vol. 2, pp. 241–260.

7 KAHN, R. L. Work and health. Chichester, New York, Brisbane, and Toronto, Wiley, 1981, p. 51.

8 LEVI, L. ed. Emotions: their parameters and measurement. New York, Raven Press, 1975.

9 LEVI, L. Psychosocial factors in preventive medicine. In: Healthy people. The Surgeon General's report on health promotion and disease prevention: background papers. Washington, DC, United States Government Printing Office, 1979, pp. 207–253 (DEHW Publication No. (PHS) 79-55071A).

10 LEVI, L., ed. Society, stress and disease: working Life. Oxford, New York, and Toronto, Oxford University Press, Vol. 4, 1981.

11 LEVI, L. Preventing work stress. Reading, MA, Addison-Wesley, 1981.

12 LEVI, L. Stress in industry: causes, effects, and prevention. Geneva, International Labour Office, 1984 (Occupational Safety and Health Series, No. 51).

13 LEVI, L. & ANDERSSON, L. Population, environment and quality of life: a contribution [of the Swedish Preparatory Committee] to the World Population Conference. Stockholm, Royal Ministry for Foreign Affairs, 1974. Also published in: Narodo-naselenije, okružajuščaja sreda i kačestvo žizni. Moscow, Ékonomika, 1979 (in Russian).

14 LEVI, L. ET AL. Work stress related to social structures and processes. In: Elliott, G. R. & Eisdorfer, C., ed. Stress and human health: analysis of implications of research. New York, Springer, 1982, pp. 119–146.

15 McLEAN, A. ET AL. ed. Reducing occupational stress. Washington, DC, United States Government Printing Office, 1978 (DEHW Publication No. (NIOSH) 78–140).

16 PINNEAU, S. R., JR. Effects of social support on occupational stresses and strains. Paper presented at the Symposium on Job Demands and Workers Health, 84th Annual Convention of the American Psychological Association, Washington, DC, September 1976. Washington, DC, American Psychological Association, 1976.

17 SWEDEN. Co-determination law The Swedish Code of Statutes, 1976 (No. 580).

18  Sweden. Work Environment Law. *The Swedish Code of Statutes*, 1977 (No. 1160).

19  Sweden. National Board of Health and Welfare. *Psykisk hälsövard—forskning, social rapportering, dokumentation och information.* [*Mental health protection and promotion—research, monitoring, documentation and information.*] Stockholm, Liber förlag, 1978 (in Swedish).

20  Sweden. National Board of Occupational Safety and Health. *Psykiska och sociala aspekter på arbetsmiljön* [*Mental and social aspects of the work environment.*] Stockholm, 1980 (document AFS 1980: 14) (in Swedish).

21  Sweden. Trade Union Confederation. [*Mental and social hazards to health in the working environment. Programme of action.*] Stockholm, 1980 (in Swedish).

22  Trist, E. L. Work improvement and industrial democracy. In: *Proceedings of the Conference on Work Organization, Technical Development and Motivation of the Individual*, Brussels, 5–7 November 1974. Luxembourg, Office for Official Publications of the Commission of the European Communities, 1974, pp. 29–61.

23  Walker, K. F. Improvement of working conditions: the role of industrial democracy. In: *Proceedings of the Conference on Work Organization, Technical Development and Motivation of the Individual*, Brussels, 5–7 November 1974. Luxembourg, Office for Official Publications of the Commission of the European Communities, 1974, pp. 83–97.

24  World Health Organization. Constitution of the World Health Organization. In: *Basic documents*, Thirty-sixth edition, 1986, p. 1.

# Chapter 17

# Coping with stress in organizations: the role of management

Cary L. Cooper[1]

## Introduction

People, the most important resource an organization can possess, should command a great deal of attention from management but frequently do not. In the words of one author musing on the implications of treating people as another form of capital asset, "Salaries and benefits are really regarded as maintenance expenses—something to be kept as low as possible as long as the machine does not break down. There is no capital cost and therefore no need for depreciation. Indeed, the return on investment of most companies would look very strange if their human assets were capitalized at, say, ten times their annual maintenance cost, and depreciated over 20 years"(10).

The costs to industry of stress-related illness alone are staggering. In 1976, the American Heart Association estimated that the cost of stress-induced cardiovascular disease in the USA was US$ 26 700 million a year. It was calculated that in the United Kingdom industry loses millions of working days a year through absenteeism due to short-term stress-related illness, at an estimated cost of £55 million in national insurance and social security payments alone. It was estimated that, in the USA, the pressures of work create physical, social, and psychosocial problems that may cost in the region of 1–3% of the gross national product (12).

If personnel officers and cost accountants in organizations were to focus on the financial costs of their human assets, it might be possible to pursue more flexible, imaginative, and futuristic personnel policies. At present, corporate planners can choose not to concern themselves with this fickle piece of human "machinery", discounting or depreciating it at will. If they were to be encouraged to answer a series of questions such as that given below they might become more aware of the value of their human resources:

—What is the total value of your organization's human assets?

—Is it appreciating, remaining constant, or being depleted?

—How much money was spent last year on the recruitment and selection of personnel?

—Was that expenditure worthwhile?

[1] Department of Management Sciences, University of Manchester, Manchester, England.

— Does your organization have data on the standard costs of recruitment, selection, and placement in order to prepare manpower budgets and to control personnel costs?

— Were the actual costs incurred last year less than, equal to, or greater than standard personnel acquisition and placement costs?

— How much money was spent last year on training and developing staff?

— What was the return on your investment in training and development?

— How does that return compare with alternative investment opportunities?

— How much human capital was lost to illness or premature death?

— How much does it cost to replace those people?

— How many young people did you lose because of your promotion and/or mobility policy?

— What was the cost of losing those people in terms of potential expertise?

— How many women have you employed, and what has been the cost of their turnover in comparison with their male counterparts?

— Does the organization really reward managers for increasing the value of their subordinates to the organization?

— Does your promotion system accurately reflect the managers' value to the organization?

— Does your firm assess in quantitative terms the effects of corporate strategies on its human resources?

Corporate planners and personnel policy makers must begin to ask themselves this type of question over the next decade if they are to make rational decisions about selection, training, and career development for the best possible use of human resources. The long-term gains could be as beneficial to the organization as the microchip or any other new type of technological development.

The purpose of this chapter is to stimulate organizations to think about their policies towards workers and their families, so that they can make better and more humane use of one of their most precious assets. Managers and other corporate planners seem to have learnt a great deal from the so-called Volvo effect. In the early 1970s, when job redesign and the creation of autonomous work groups were considered to be the answer to all industrial ills, many people were influenced by the publicity given by the media to the experiment undertaken by Volvo motor works. A new plant was designed in collaboration with the workers at Kalmar in Sweden, to meet their personal and work-related needs. It provided facilities for a production system that enabled one small group of workers to assemble a complete car from start to finish. Other companies, provided with an example of what was hailed as the latest management technique, unfortunately tried to introduce it without critically evaluating it in the context of their own organizations and adapting it as necessary.

## What Western Europeans think

To illustrate the changing attitudes and values of managers towards the interface of work and family, it is not necessary to look further than a large-scale survey carried out by the International Management Association (2). Three thousand middle to senior executives in ten Western European countries were questioned on their life values. In almost every country what gave them the most satisfaction proved to be home life (Table 4). Only a third said their careers offered the most satisfaction.

Asked whether anxiety about their jobs frequently spilled over into their home lives—the executives in many of the countries

Table 4.  Survey of executives in 10 countries[a]
Question: What gives you the most satisfaction?

|  | Home life % | Outside interests % | Career % |
|---|---|---|---|
| United Kingdom of Great Britain and Northern Ireland | 50.0 | 13.3 | 35.0 |
| Denmark | 44.8 | 9.4 | 39.6 |
| Switzerland | 51.9 | 10.2 | 32.4 |
| France | 55.8 | 15.1 | 26.7 |
| Germany, Federal Republic of | 45.0 | 7.5 | 42.5 |
| Sweden | 47.5 | 8.9 | 36.6 |
| Italy | 52.2 | 21.7 | 21.7 |
| Belgium | 48.9 | 18.1 | 33.0 |
| Spain | 57.0 | 16.1 | 24.7 |
| Netherlands | 36.9 | 20.2 | 35.7 |
| Total | 49.0 | 14.0 | 32.8 |

[a] Reprinted by special permission of McGraw-Hill International Publications Company Limited, London, England. All rights reserved. Copyright (c). (*International management*, **35**(7): 12–20 (1980)).

perceived that problems at work were affecting their home environments; and the figures shown in Table 5 are probably low estimates, since many executives are unaware, or choose not to be aware, of how life at work affects life at home.

Table 5.   Survey of executives in 10 countries[a]
Question: Does your anxiety about your job frequently spill over into your home life?

| | Yes % | No % |
|---|---|---|
| United Kingdom of Great Britain and Northern Ireland | 40.5 | 58.6 |
| Denmark | 33.3 | 65.6 |
| Switzerland | 30.4 | 68.6 |
| France | 47.0 | 50.6 |
| Germany, Federal Republic of | 29.9 | 67.5 |
| Sweden | 29.0 | 71.0 |
| Italy | 46.2 | 50.8 |
| Belgium | 29.7 | 70.3 |
| Spain | 41.7 | 54.8 |
| Netherlands | 25.0 | 70.0 |
| Total | 35.3 | 62.8 |

[a] Reprinted by special permission of McGraw-Hill International Publications Company Limited, London, England. All rights reserved. Copyright (c). (*International management*, **35**(7): 12–20 (1980)).

Table 6.   Survey of executives in 10 countries[a]
Question: Would you give up an important function at home if it conflicted with an important job-related function?

| | Yes % | No % |
|---|---|---|
| United Kingdom of Great Britain and Northern Ireland | 63.8 (82.8) [b] | 34.5 (13.8) |
| Denmark | 60.2 (73.1) | 30.1 (20.4) |
| Switzerland | 60.8 (70.6) | 36.3 (25.5) |
| France | 67.5 (63.9) | 24.1 (27.7) |
| Germany, Federal Republic of | 71.4 (76.6) | 24.7 (23.4) |
| Sweden | 57.0 (61.0) | 38.0 (32.0) |
| Italy | 53.8 (53.8) | 38.5 (38.5) |
| Belgium | 68.1 (68.1) | 28.6 (29.7) |
| Spain | 60.7 (58.3) | 36.9 (39.3) |
| Netherlands | 53.8 (58.8) | 37.5 (33.8) |
| Total | 61.7 (66.7) | 32.9 (28.4) |

[a] Reprinted by special permission of McGraw-Hill International Publications Company Limited, London, England. All rights reserved. Copyright (c). (*International management*, **35**(7): 12–20 (1980)).
[b] Figures in brackets represent the choice the executives said they would have made five years ago.

Table 7.   Survey of executives in 10 countries[a]
Question: To further your career, would you uproot your family now to move to a new location for a higher paying and more responsible job?

|  | Yes<br>% | No<br>% |
|---|---|---|
| United Kingdom of<br>  Great Britain and Northern Ireland | 56.0 (70.0)[b] | 43.1 (29.3) |
| Denmark | 28.0 (62.4) | 66.7 (33.3) |
| Switzerland | 49.0 (74.5) | 50.0 (24.5) |
| France | 51.8 (68.7) | 42.2 (25.3) |
| Germany, Federal Republic of | 53.2 (72.7) | 45.5 (27.5) |
| Sweden | 38.0 (71.0) | 59.0 (28.0) |
| Italy | 55.4 (70.8) | 38.5 (23.1) |
| Belgium | 48.4 (60.4) | 47.3 (37.4) |
| Spain | 50.0 (73.8) | 45.2 (22.6) |
| Netherlands | 43.8 (73.8) | 53.8 (25.0) |
| Total | 47.4 (69.8) | 49.1 (27.6) |

[a] Reprinted by special permission of McGraw-Hill International Publications Company Limited, London, England. All rights reserved. Copyright (c). (*International management*, **35**(7): 12–20 (1980)).
[b] Figures in brackets represent the choice the executives said they would have made five years ago.

Very interesting results came from questions about specific home/work interface areas—i.e., relocation, and conflicting priorities at home and at work. The executives were asked if they would give up attending an important function at home if it conflicted with an important work-related function.

Although a majority said they would still attend the important event at work, there was a discernable trend towards favouring family commitments compared with what they said they would have done five years ago. In Denmark, the Federal Republic of Germany, Switzerland, and the United Kingdom this trend was quite noticeable. When asked whether, to further their careers, they would uproot their families to move to a new location for a higher paying and more responsible job, the majority responded that they would not. It is revealing that five years ago nearly 70 % would have accepted the promotion even though it entailed a move, whereas now only 47 % would be prepared to accept it. The reversal of attitudes was most dramatic in Denmark, the Netherlands, Sweden, and Switzerland.

The gap between working life and home life is often perceived as being too large. The survey also showed that the number of dual career managerial families is increasing, a fact often ignored.

With these changes in mind, what can be done by the organization to support the executive family?

## The corporate wife

It is obvious that one of the main sources of stress in the executive family is the fact that managers and their wives know very little about what is going on in each other's lives. Many organizations believe that the corporate wife will be placated if she is allowed to accompany her husband to a company conference at a sumptuous and idyllic retreat, and that this is an adequate basis for establishing a communications link. It has, however, been suggested that "The wives are pushed into buses and vanish for the day. Then they reappear to freshen up for a nice dinner and dance in the evening. The wives remain totally detached from the conference. They know nothing of what has been going on. What was the purpose of inviting them along in the first place?" (1).

Most organizations would find it difficult to answer this question; some are trying to solve the basic problem by, for example, sending male managers and their wives on seminars at a place away from the scene of their working lives where they attend lectures on such topics as "the changing role of husband and wife in family life", "the problems of corporate wives", "the problems of managing in the 1980s", "coping with stress", "progressive education", "youth employment problems", "alcoholism", and "company loyalty" (1). The couples are encouraged to discuss their personal and family dilemmas and problems in small groups and, by sharing them with others in similar circumstances, they can begin to see that they suffer the same difficulties and to share each other's coping strategies.

This is an approach in the right direction, although there are some reservations about courses run by an outside agency for managers from different organizations. For several reasons it is preferable for this type of programme to be arranged within the organization; not only would work/home interface tensions thus be reduced, but also the creation of a more open climate in the organization itself might be stimulated. Given the current, rather less than open and consultative, nature of most organizations, this type of approach might be difficult to initiate, but once norms were changed to accommodate it, the benefits would more than offset the initial unease.

In addition to job awareness seminars, a few organizations are considering more radical alternatives, such as paying for wives to accompany their husbands on business trips, or providing child-minding services to allow wives to engage in personal growth activities or to take up part-time or full-time employment.

## The executive woman

There is no shortage of advice on the various ways in which the female manager can help herself, both at work and in organizing her family life. But the answer to the question "Can women achieve success alone without the right corporate climate?" is a resounding "No!" Organizations must create the appropriate corporate facilities and attitudes for the executive woman to do more than just survive.

### Helping the working mother

One way in which an organization can help the executive woman is to recognize the sources of her pressures and provide support to alleviate them. A large life insurance company in the USA recognized in the early 1970s that 70% of its female employees who became pregnant left the company for good. A programme was planned to help those who wanted to work and raise a family simultaneously and by 1977 only 30% of the pregnant women were stopping work. A series of seminars was devised for over 2300 working mothers and male managers, whose wives had full-time jobs, during which a wide variety of the problems experienced by dual-career parents was explored— e.g., guilt feelings about not assuming the traditional mothering role at home, which many executive women experience.

Four of the most important lessons learnt from this programme were specified by the academic management consultant involved, namely (6):

—recognize your own needs;
—make the most of the time you do spend with the children;
—get the children involved in household tasks;
—agree on a fair division of household chores.

Helping the working mother must also entail a change in the personnel policies of most companies. To organize a training programme is one course of action, but to acknowledge the reality that dual-career managerial families exist and to accommodate them is another. Policy changes to institute the following are what is really needed:

(1) *More flexible working weeks.* This will enable work and home commitments to be coordinated. This might mean more part-time posts or merely flexible working schedules.

(2) *Paternity as well as maternity leave.* With the increase in the number of dual-career managerial families, more considerate and flexible policies will have to be introduced to enable women to continue to work when they have children. Paternity leave is essential to this process.

(3) *Day-nursery facilities.* Throughout Europe, organizations are beginning to provide in-house day-care centres for the children of their employees. Since governments are not assuming this responsibility, the organizations themselves will have to, at least in the short term.

(4) *Changes in regard to relocation.* This will involve two types of change; the first to allow women managers opportunities for promotion without moving; the second to accommodate the needs of dual-career managerial families when one spouse is offered a move and the other is not. Men should have the same right to refuse as women, to maintain the integrity of their working and home life.

## Providing career opportunities for female managers

In addition to helping the woman in a dual-career family to cope with her domestic and working environments, it is important for organizations to encourage women to enter management and to provide them with career opportunities once they are there. A number of organizations are appointing special personnel whose task is to try and ensure that the spirit of equal opportunities legislation is adhered to; in some organizations in the USA, for example, "affirmative action officers" are responsible for making sure minority groups of workers—e.g., women or blacks—are not discriminated against in such areas as hiring, promotion, or training. In addition to such obvious positive action, the whole infrastructure of an organization would have to change once the woman took up her post. The types of organizational change needed to encourage women to sustain their careers as managers would relate to (8):

— career planning and counselling;
— a decision, on the part of senior management, to promote the careers of women;

—helping male managers to come to terms with female managers;

—creating informal support networks for all female managers.

There is a variety of other ways of getting women to think about their careers. One of the most interesting is the organization of career planning workshops by a large drug company for its male and female employees. At these workshops experiential techniques are used in small groups of four to six (4). The members of each group are encouraged to talk and are asked to write down a word or phrase they feel is descriptive of who or what they are, using nouns instead of adjectives—e.g., woman, mother, lawyer—and to rank them in order of preferred priority. They are also asked to draw life and career graphs; that is charts showing the most significant events at the high and low points of their personal lives and careers. Most important of all from a career-planning point of view, they are asked to extend the line five years and ten years into the future. They are then encouraged to write brief descriptions of their ideal jobs, taking into consideration their experience, skills, and preferred priorities. Group discussions are encouraged at each phase of the workshop programme. By this means the individual is helped to concentrate on the past, the present, and future prospects, in a way that can help in career planning.

## Forcing change

To the debate on what organizations can do to help their female managers might be added the fact that some are not content to wait for change but are forcing their male managers to adapt to modern thinking by instituting hiring and promotion policies that call for immediate action to include women. For example, a large food company in the USA adopted a policy providing bonus payments to its managers if they could increase the numbers of minority members employed in the organization. Bonus payments can be as high as 30 % of the basic salary. Each branch of the company is obliged to set itself short- and long-term equal-employment goals, based on official population census statistics for the particular town in which it is sited. From that figure the chief executive of the branch determines what proportion of the various job categories available should be filled by women. If this proportion is achieved or exceeded, he is awarded a commensurate "social responsibility bonus".

It is obvious that this approach to the employment of women and other minorities is very rare in industry, but in the short and medium term it is the kind of corporate policy that might be necessary if moral obligations to those who have for so long been denied equal opportunities are to be fulfilled.

## Managerial relocation

### A survey of company policy

A major source of pressure for the executive and his family is frequent moving (4). Although, in many cases, he appears to benefit materially from such relocation—in terms of an increase in salary through promotion, new furnishings, an increase in assistance with a mortgage, etc.—the psychological and social costs to the members of the family are great. In the United Kingdom, for example, it is common for managers to move on average every two or three years until the age of 50 years.

This state of affairs creates a variety of problems for the executive family. Firstly, there are the arrangements to be made in regard to housing. Secondly, there is the strain imposed while the executive is a "weekend father", during the inevitable period when he is working in one part of the country in his new job and the family remains in another. Incidentally, this need not be a problem if companies would only allow the time for the manager and his family to move together as a unit; in most cases relocation could be organized in such a way, since the services of the executive to be transferred are seldom required immediately. Thirdly, moving children from one school to another creates great difficulties in terms of their academic and social development and the emotional energy of the family. Finally, the constant moving of an executive, around the country or abroad, creates an obstacle if his wife has a career. There is a variety of other problems associated with moves but these are some of the reasons why so many managers are refusing moves, and consequently promotion, since to refuse often means no further promotion. There is no better way of creating job demotivation and dissatisfaction.

A recent survey conducted by Merrill Lynch Relocation Management in the USA shows that corporate attitudes and policies towards the relocation of managers may be changing (16). Companies are helping managers to transfer their homes. When they are moved to a part of the country where the cost of living is high they are given differential payments. Some corporations are also changing their attitudes and policies towards the working spouses of relocated managers, helping them to find jobs. Some organizations are beginning to develop more liberal attitudes towards refusals to move.

The vast majority of middle-sized United States companies and almost all European companies of any size are lagging behind on most of the above-mentioned policies. This means not only a likely increase in stress-related problems and intolerable family strains, but also that one of the most important social trends for the next two decades, the event of the dual-career family—which will exacerbate the problems—is being ignored. The changes in social institutions—e.g., dual-career marriages, shared parenting, etc.—mean that organizations must begin to look beyond merely establishing temporary support systems and consider the wider issues.

They could consider coordinating their relocation plans with the phases of the employee's life cycle outside work. Obviously at certain times in an employee's family life cycle change would be less disruptive—e.g., when he is single rather than when he is married and the children are beginning school or his wife's career is developing. There are occasions, of course, when a move is inevitable and necessary, such as when a manager possesses the skills vital to a particular project.

The wife's role in the development of her husband's career has been virtually ignored. It would seem reasonable that she should be given the option of getting involved in the making of decisions with regard to a move. At present, organizations are contracting with one element of the family but making decisions that radically affect the family as a whole.

Organizations frequently give their managers inadequate notice of a move, thus involving them in a period of separation from the family, which may adversely affect them, the family, and, in the long run, the organization, which may suffer from less than efficient work performance as a direct consequence of the domestic conflict. Notice of a relocation should be given long enough in advance to minimize the period of separation from the family. On the rare occasion when it may not be possible to give adequate warning, consideration could be given to:

—making time off to find housing and to carry out the move the norm rather than a privilege;
—allowing business travel to be kept to a minimum so that frequent contact can be maintained between manager and family, for example, at weekends;

—making formal or informal arrangements for the spouse left with the responsibility of the home and children to have someone to turn to in a domestic emergency during the separation;

—helping managers and their spouses to anticipate the problems that might be encountered during the separation;

—allowing work on the new responsibilities to be carried out from home, or the previous office.

In some circumstances separation could cause an individual or a family relationship to grow, as was indicated by some respondents in a study published in 1976 (13), but it is important that the manager and his wife should be allowed to have some influence on the decision regarding the length of separation.

Misleading or imprecise information given by organizations to their managers about what is planned for their careers nurtures uncertainty and is a major source of the stress involved in a potential move. If up-to-date, honest information is provided about prospective plans, uncertainty is bound to be minimized leading to greater acceptance and smoother transition if they come to fruition. As far as possible a manager should have the right to refuse to accept plans that might have harmful consequences for himself or his family, without feeling, or being made to feel, that his career would suffer unreasonably.

Organizations will have to reconsider their attitudes to women executives, offering them moves that involve promotion on the same basis as men, allowing them the same opportunities to refuse a transfer without damaging their promotional prospects, and providing them with the same support facilities. This is a time of great social change; corporate planners will have to use insight and be creative and innovative in designing effective strategies to deal with it.

After having moved a manager, an organization's responsibility for him and his family is not over. There are a number of issues related to the move that might be considered. The organization can help to minimize the adjustment difficulties faced by the wife and children in their new environment. Many wives feel isolated and alone for several months after entering a new community (13, 17). In addition, they have to cope with the adjustment problems of their children at new schools and in new peer groups. There are no universally acceptable solutions to these problems but consideration might be given to:

—discussing with the wife and other members of the family the possible difficulties they might encounter;

— providing the wife with opportunities, such as social occasions or discussion groups, to meet other company wives under the organization's auspices;

— providing help in the house for the first few months so that the wife has free time to get to know members of the community, find a job, or find some activities that will give support to the children.

These and similar approaches might help the family, especially the wife, to settle down more easily. They would also be a means of providing visible support at an emotionally trying time. The organization should also be aware of the pressures that newly arrived managers may be under during their first months at a new site.

## The managerial sabbatical

A way of providing a manager with the experience of working in a different environment, with people with different interests and approaches to problems, is the sabbatical system.

Table 8 shows the types of sabbatical that would help managers to develop their careers, and that could be flexibly organized to allow for individual needs and circumstances (11).

## Improving the health of managers

An increasing number of organizations are providing health and counselling facilities to their executives, because of the appalling statistics on stress-related illness and coronary heart disease among younger managers, and because of the costs of not protecting them. Costs relate to the days lost to illness, lost expertise, and, more recently, a massive increase in worker compensation claims for, what are said to be, the results of stress at work. In California, for example, the courts have been upholding "cumulative stress or trauma" claims against employers who have not been trying to minimize the pressures of work; the statistics of one of the largest industrial insurance companies showed a large increase in such claims during the 1970s. Cumulative trauma is the term given to a type of compensation claim in which a worker contends that a major illness or disability is the cumulative result of minor job stresses and strains over a period of years. This type of claim became popular in the early 1970s and the fact that many were

Table 8.  Types of managerial sabbatical and the potential benefits[a]

| Type | Benefits |
|---|---|
| To a university or an institute of higher learning | Opportunity to view career and work from a distance<br>Opportunity to study and update knowledge in a particular field<br>Opportunity to apply theory to practice on the basis of work experience |
| To a government or quasi-government agency | Opportunity to apply skills to national problems<br>Opportunity to work with those at the power base of a country |
| To a management consultancy | Opportunity to work in another company and learn how it operates |
| To a company in the same or a related industry | Opportunity to compare current knowledge and skills and develop new ones |
| To a client company | Opportunity to see how own company's products are utilized<br>Opportunity to assist the client company in the development of new applications of own company's products |
| To other functions | Opportunity to broaden management experience |
| From the headquarters of the company to a branch, or vice versa | Opportunity to learn how others in the organization operate |

[a] Adapted from reference 11.

successful was due to the inability of the industrial organizations concerned to show that they had taken adequate measures to minimize the stress-related characteristics of various jobs or to provide counselling or other stress-coping facilities at the place of work.

## Keep-fit programmes

Many USA and some European organizations have begun to provide extensive keep-fit facilities for their managers (5). In Canada two life insurance companies participated in a joint research project to see what the effects of a keep-fit programme would have on their managers. In all 1125 managers were enrolled in a systematic physical fitness course held at the headquarters gymnasium. The results were interesting; absenteeism dropped by 22%, and productivity in the exercising group rose by 3%. In addition, the managers enrolled in the course developed a significantly more positive attitude towards their work and reported better relationships with their superiors and subordinates.

A medium-sized company on the east coast of the USA introduced a variation of the keep-fit idea by offering its employees a voluntary 12-week programme consisting of a relaxation training break during normal working hours. Over 140 employees who volunteered to take part in the relaxation training break were compared with 63 employees who did not take part, who were selected randomly. The volunteers agreed to keep daily records for 12 weeks and to have their blood pressure measured. In addition, their general health and their job performance were assessed during the period of the experiment. The results indicated that not only was a relaxation training break feasible within normal working hours but it led to improvements in general health, job performance, and wellbeing, as well as significantly decreasing blood pressure throughout the entire period.

More detailed examples of health care programmes can be found in a book of case studies on stress prevention approaches by different companies (*14*).

## Stress counselling

Occasionally, an organization employs a professional psychologist to discuss personal problems with its employees. Such a post could be extremely helpful in assisting individuals to find means of mitigating stress. The problem lies in their reluctance to be seen seeking assistance. It is thus sometimes more effective if the expert can masquerade under the title of welfare officer or management development officer, allowing advice to be sought without necessarily being seen to acknowledge a personal psychological impairment. For the long-term benefit of the organization, however, it is preferable to acknowledge the existence of stress openly and to establish a counselling service.

A similar type of function could be assumed by the incumbents of at least two other professional posts existing in some organizations: the physician and, when appropriate, the industrial chaplain. Both have the great advantage of being perceived as independent of the organization and its decision-making process. Both are regarded as repositories of confidences and as sources of two kinds of solace—medication and spiritual guidance—that the organization cannot provide.

If counselling on stress is given by the physician formally on demand, it may well produce the type of inhibition already described but an established practice of regularly checking executives' health sets up a cycle of encounters that allows for the possibility of dircussing stress and its effects. There are additional obvious advantages in that a physician can diagnose a wider range of stress symptoms than another potential intervener and can infer a proneness to stress from a medical history.

## Training managers to be aware of stress in others

The poet, W. H. Auden, wrote in *The Unknown Citizen*,
"Was he free? Was he happy? The question is absurd:
Had anything been wrong, we should certainly have heard."
But can we, or should we, rely on informal hearsay to judge the wellbeing of a fellow employee?

Stressful behaviour can all too frequently be disguised or covered up so that it is acceptable, at least in its initial stages. It is, therefore, important, with the increasing pressures of life in an organization, to make managers aware of stress manifestation in themselves and in others, so that they are in a better position to act. Those showing signs of stress can be encouraged to participate in a stress reduction programme, or the job or workplace factor responsible can be modified to alleviate the effect of the stressor.

A 1–2-day training programme can easily be devised to focus on the physical, emotional, and behavioural symptoms that may be stress-related, such as those, shown in Table 9, found during a 10-year longitudinal study of 500 senior executives in the United Kingdom (7, 15).

The training programme could also include sessions in which the executives' experiences in reacting to various pressures are analysed.

## Bridging the gap between work and home

The gap between work and home could be bridged in a variety of ways. Obviously a most important first step is to find out from the managers and their families where the areas of conflict and difficulty lie. This could be done by extending the regular surveys carried out in some companies to find out managers' attitudes to company practices and policies. A Scandinavian oil company, for example, conducts such a survey each year.

Table 9. Physical, emotional, and behavioural symptoms of stress in managers

**Physical**

Palpitation—an awareness that the heart is beating forcefully, irregularly, or quickly
Pain and tightness in the chest
Indigestion and gaseous abdominal distension
Spasmodic, griping abdominal pain and diarrhoea
Frequent urination
Impotence or lack of libido
Alteration of the menstrual pattern
Tingling feelings in the arms and legs
Muscle tension—often pain in the neck or lower part of the back
Persistent headache—often starting in the neck and extending forward over the head
Migraine
Skin rash
Feeling of a lump in the throat
Double vision and difficulty in focusing the eyes

**Emotional**

Excessive and rapid swings in mood
Worrying unreasonably about things that do not matter
Inability to feel sympathy for other people
Excessive concern about physical health
Withdrawing and day-dreaming
Feeling tired and lacking concentration
Increased irritability and anxiety

**Behavioural**

Indecision and unreasonable complaints
Increased absenteeism and delayed recovery from accidents and illness
Accident-proneness and careless driving
Poor work, cheating, and evasion
Increased smoking
Increased consumption of alcohol
Increased dependence on drugs—tranquillizers and sleeping tablets
Over-eating or, less commonly, loss of appetite
Changed sleeping pattern—difficulty getting to sleep and waking tired
Impaired quality and quantity of work

Some of the more obvious difficulties could be minimized in the ways described below.

## Business travel

Travel abroad on business sometimes helps to widen the gap between manager and family. An organization could help to partially overcome the effects of such absences by allowing the spouse to accompany the manager on a certain number of business trips each year. A large number of the executives questioned during the International Management Association survey felt that this would be a very desirable policy (2).

Organizations could do more to help executives who have often to travel long distances by air by providing them with more time to prepare for each trip and to recover from it, and with information about what to do or not to do before, during, and after a long flight. For example, the overseas medical officer of a European national carrier suggests the following 10-point plan:

—plan the flight well in advance, arranging it so that you arrive at your destination preferably around your usual bedtime;
—try to relax during the 24 hours before a long flight;
—reduce the amount you smoke before and during a flight;
—do not indulge heavily in alcohol during the flight;
—drink plenty of non-alcoholic fluids during the flight;
—avoid eating fried or fatty foods;
—wear loose-fitting, comfortable clothes and shoes;
—carry light but warm clothing;
—take a 24-hour rest after a long flight; do not go straight into a business meeting;
—carry mild laxatives and sedatives if travelling long distances.

## The business lunch

Many companies encourage their executives to meet clients for lunch, or arrange lunches for internal organizational purposes. Clients and employees are often taken out to first class restaurants. On many occasions the business lunch may not be necessary although, on some, it may serve the purpose of providing the individual manager with a reward for service or a job well done. This is a practice that is important in terms of the quality of life as a manager but a heavy meal in the middle of the day, with a high energy, high cholesterol, and high alcohol intake, cannot do the health of someone with a sedentary job very much good. A possible solution might be for companies to scrutinize and establish the real purpose of the so-called business lunch. If it is used as a means of rewarding executives some other method should be found. An organization's most innocent-appearing activity can create conflict within a family and an awareness of this fact might help to minimize the conflict.

## Long hours

All too often long hours become the norm; executives are made to feel guilty if they leave at 17 h 00 and they are subtly encouraged to work into the early evening. This can dramatically affect the activities of the family, reducing the "quality time" the manager can spend within it. It should be asked whether it is necessary to work 60 hours a week; what the organization is really saying when it encourages this as the norm; whether it is merely a mechanism to obtain a show of organizational commitment; and, if so, whether there are not better ways of doing this than creating unnecessary home versus work conflict. Indeed, it would seem that long hours might have the opposite effect, creating antagonism towards the organization and what it represents; making it seem the great intruder in the family home; the unfeeling, uncaring employer who takes a parent or spouse away from family activities at will. The questions to be asked in regard to individual managers are:

— is it necessary to work beyond a 40-hour week and, if so, why?
— how are the long hours affecting work and family life? If they are adversely affecting both, what can be done to minimize the effects or what support can be provided?
— could the working hours be made more flexible to allow "quality time" to be spent at home during busy periods or projects?

## Support systems for managers

If the norms within a company can be changed to acknowledge and accept the reality of stress in organizational life, the process of managing it becomes easier and probably more effective. This is because one of the most efficacious ways of preventing, rather than coping with, the pressures of working in an organization is to establish a support system among the managerial staff: that is to encourage the managers to help one another and share their concerns, to reorganize work schedules and tasks according to individual needs, to acknowledge that family circumstances can affect performance, and to counsel or seek help for their colleagues, subordinate managers, or superiors if they are experiencing problems at home. A management development adviser could help to create a support group. Such a group should not be a mechanism that simply deals with problems as they arise or with managers who are manifestly at the end of their tether, but rather should meet regularly so that problems and potential stressors can be anticipated and individuals helped to perceive them and to plan their activities in a way that will minimize or eliminate them before it becomes too late.

This raises the question of competition among managers and whether it is necessary or healthy for an organization. It was found in a study at the University of Texas that the most competitive managers had not necessarily achieved the most (3). Success was much more the result of a desire to work hard and the possession of a drive to achieve personal standards of excellence than of aggressive or competitive behaviour. It is being postulated more and more that competitive and cut-throat behaviour at work not only leads to stress-related illness and job dissatisfaction but also to lower overall organizational productivity.

## References

1 ARBOSE, J. Involving the corporate wife. *International management,* **34**(10): 27–28 (1979).

2 ARBOSE, J. The changing life values of today's executives. *International management,* **35**(7): 12–20 (1980).

3 BENSAHEL, J. G. Why competition may not always be health. *International management,* **33**:(10): 23–25 (1978).

4 COOPER, C. L. *The executive gypsy.* London, Macmillan, 1979.

5 COOPER, C. L. *The stress check.* Englewood Cliffs, NJ, Prentice-Hall, 1980.

6 COOPER, C. L. *Executive families under stress.* London, Macmillan, 1981.

7 COOPER, C. L. & MELHUISH, A. Occupational stress and managers. *Journal of occupational medicine,* **22**: 588–592 (1980).

8 EKBERG-JORDAN, S. Preparing for the future: commitment and action. *Atlanta economic review,* March 1976. pp. 47–49.

9 FLAMHOLTZ, E. Should your organization attempt to value its human resources management. *California review,* March, 1971.

10 HANDY, C. *Understanding organizations.* London, Penguin, 1976.

11 JONES, A. N. & COOPER, C. L. *Combating managerial obsolescence.* Oxford, Philip Allan, 1980.

12 McMURRAY, R. N. Mental illness: society's and industry's six billion dollar burden. In: Noland, R. L., ed. *Industrial mental health and employee counselling.* New York, Human Sciences Press, 1973.

13 MARSHALL, J. & COOPER, C. L. *The mobile manager and his wife*. Bradford, MCB Publications, 1976.

14 MARSHALL, J. & COOPER, C. L. *Coping with stress at work: case studies in industry*. Epping, Essex, Gower Press, 1982.

15 MELHUISH, A. *Executive health*. London, Business Books, 1978.

16 Merrill Lynch Relocation Management. *A study of employee relocation policies among major US corporations*. New York, 1980.

17 SEIDENBERG, R. *Corporate wives-corporate casualties?* New York, AMACOM, Division of American Management Association, 1973.

Chapter 18
# Controlling physiological stress reactions
H. Reginald Beech[1]

The requirements of a scientific approach in psychology have been succintly reviewed in the past, and psychotherapy has been contrasted with the behavioural approach (7). In general terms it may be said that the behavioural approach is characterized by its emphasis on properly formulated theories and testabhe deductions, exemplified in experimental work with appropriate controls. In particular, it accords special places to the role of learning in symptom formation, the conditioning effect of symptoms, and a contemporaneous rather than historical perspective of dysfunction. In short, although the behavioural approach is a possible theoretical framework within which to view psychological function and dysfunction, it is scientific in character, and is backed by a vast store of empirical data on human performance under many different conditions. That such an approach should have ready application to the field of stress is understandable (3, 5).

## Types of stressor

One way of looking at occupational stress is to begin with the proposition that it results from a discrepancy between workload and capacity, from a failure of work to satisfy certain needs, or from a perceived threat to capacity or satisfaction. In these terms certain types of stressor, such as work overload when, for one reason or another, the individual is unable to carry out his assignments in the time set aside for them, are readily appreciated. A further familiar example of overload is found when the task to which the individual is committed is beyond his technical or intellectual competence.

Stressors in a second category arise from frustrations, among which are those related to uncertainties about job responsibility, conflicts within the job function, poor career prospects, communication failures or difficulties, and problems arising from work structure and organization.

A third category includes those that may result from changes, in the form, for example, of tecnical innovations, work relocation, promotion, retirement, or loss of job.

Whatever the source of stress the individual's style of reaction will tend to be idiosyncratic rather than particularly related to

[1] Department of Psychiatry, The University Hospital of South Manchester, Manchester, England.

the type of stressor involved. Such reactions fall into two main categories: those that are physical/physiological in character and those that are of a more psychological nature. Numerous influences are undoubtedly at work in determining which of these response styles is displayed, or the relative weight given to each when both are exhibited. When psychological reactions are involved, however, congruity is generally found between cognitions (thinking), emotions (feelings), and behaviour. For convenience, however, the three are considered independently.

## Cognitions

What the individual thinks of himself and the world in which he lives is quite obviously central to his adjustment. If he perceives the world to be hostile and unfriendly, or that others do not appreciate him, his feelings and actions will be influenced accordingly. Such cognitions can take many forms and characters, and each individual has his own unique store of constructs about himself and his world. Understandably, the imposition of stressors tends to activate that part of the construct store containing cognitions of a particular quality. Some will perhaps be helpful if they are present in moderation—e.g., self-assertion, a determination to overcome difficulties, or a general philosophical attitude that not every battle can be won or it is no use crying over spilt milk. Others will be less helpful, or even destructive, if they are notions reflecting exaggerated concern or fear about current or future events and the ability to meet and cope with them. It is, of course, such negative cognitions that usually become the focus of attention in those suffering from reaction to stress.

The mechanism involved is equally complex. There is little doubt that a reciprocal relationship exists between the emotional and cognitive states, and that one influences the other. Often, however, it seems that in the ensuing chain of events the primary factor becomes obscured, to such an extent that it may be pointless to inquire into the origins of the distressed condition. It is relatively easy to illustrate this complexity with an example of a situation that will be familiar to most readers.

Mr S, a middle-level manager in a large engineering company, and his wife were experiencing considerable marital difficulties and felt themselves to be drifting apart. Finally, Mrs S formed a relationship outside the marriage. For some time Mr S tried to continue with his work as before but the strain of the domestic situation made him less tolerant and more impulsive, leading eventually to severe "personality" clashes. Feeling that just about "everyone and everything" were against him, he elected to take a firmer grip on the situation, becoming autocratic, even tyrannical, at the office, in an attempt to gain control over his disintegrating world. Naturally, further problems arose from the way he behaved and he began to drink heavily to relieve the tension he was experiencing. Within a short space of time he was involved in a serious road accident and was fortunate only to be given a heavy fine and lose his driving licence. In short, the defensive strategies adopted by the individual under stress can often be seen to compound the difficulties and produce problems of an even more severe nature.

A further complicating feature, despite what has just been said about the confusion arising from a process of cause and effect, concerns the possible role of a biological alteration of state as a primary factor. Altered mood, whether towards euphoria or towards dysphoria, affects the cognitions. When mood changes in the direction of depression, ideas consistent with that state will form, whether the dysphoria is inspired by biological or psychological factors. Clearly, when a biological alteration of state has precipitated a chain reaction of the kind illustrated, its continuing existence serves to initiate further problems. If it were possible to determine such a degree of biologically altered mood state in a stress reaction, appropriate therapeutic action might be taken, perhaps pharmacological rather than psychological.

Nevertheless, in the context of occupational stress, psychological causes and psychological treatments are generally most often implicated. The basic question, in the context of the reciprocal relationship between the character of the thoughts and the emotion experienced, is whether changes in the cognitions leading directly to the desired alteration in feelings and changes in behaviour can be engineered. In an analysis of assertive behaviour no differences in awareness of what an appropriate social response should be were found between less assertive and highly assertive individuals, but a characteristic of the former was to make a negative subjective statement in a real-life situation—e.g., "I was worried what the other person would think of me"—while the latter were more likely to make a positive statement—e.g., "I was thinking that I was perfectly free to say no" (13). Research suggests that to alter cognitions can lead to useful changes in feelings and behaviour (11).

Typically, the cognitions it is sought to change are distortions of reality; they involve:

(1) *Overgeneralization*—a conclusion is drawn on the basis of relatively little data.

(2) *Selective abstraction*—detail is taken out of context and given undue emphasis.

(3) *Arbitrary inference*—unwarranted inferences are drawn from the evidence.

(4) *Magnification*—exaggerated importance is given to an event.

(5) *Dichotomous thinking*—events are misclassified by being seen as having dire significance.

(6) *Personalization*—unpleasant events are seen as carrying personal implications when there is no logical basis for such a judgement.

The lecturer who sees one or two individuals slipping out of the door at the back of the hall may engage in a distortion of this kind, concluding, "My performance is awful; they can't stand any more; I am absolutely hopeless". Negative cognitions such as this may be a part of a more generalized low self-esteem or another aberrant aspect of personal evaluation.

A means of dealing with this type of thought forms the *rational emotive therapy* proposed by Ellis (6), which basically sets out to eliminate disruptive irrational thought by:

— identifying the irrationalities;
— examining the irrationalities carefully;
— disinguishing irrational thoughts from rational thoughts;
— avoiding distortions of thinking and adhering to the realities of the situation.

Once disruptive thought patterns are challenged and changed, it is argued, the accompanying unpleasant emotions and disturbances in behaviour are eliminated and, it is hoped, the negative and exaggerated views are replaced by more constructive ideas based on reality.

Whether faulty ideas can so easily be dispelled may be doubtful; certainly, in some cases, the individual has no problem in seeing the irrationality of his thoughts but cannot prevent himself from feeling and behaving as if he is threatened. The executive who is subject to anxiety at board meetings may be clearly aware that he *is* competent, that no one threatens him, that he is appreciated and valued by other company members. Nevertheless, he experiences considerable emotional upset at company meetings.

It may be argued that such reactions have little to do with the illogicality of the thought processes but, rather, are the product of faulty learning or conditioning. Hence, the means of rectification must involve reconditioning or deconditioning. A dog, in Pavlov's experiments, salivating to the ringing of a bell as a conditional response, after experiments in which the ringing of a bell preceded food, did not do so as a result of careful—albeit erroneous—thinking about the significance of a bell ringing but as a kind of automatic or reflex response. In that context, it could be argued, an alternative approach to cognitive restructuring might be more beneficial.

One such approach involves "thought stopping", where the aim is to produce conditioned inhibition of a troublesome thought. Briefly, the capacity to inhibit an idea "at will" is developed in stages, by a process involving the deliberate evocation of the thought, followed by its active suppression on a given cue. Eventually, the cue can be one word, "stop", uttered as "internalized speech". An alternative, similar in mechanism, is to use an easily applied aversive stimulus whenever the unwanted thought occurs, such as snapping an elastic band worn on the wrist (11).

Stress inoculation is a further alternative also requiring a degree of self-management on the part of the individual, but aiming at the regulation of emotion through cognitive control, rather than suppression as in thought stopping. The process involves two main stages:

(1) *Cognitive preparation*, in which the individual learns more about emotion and its determinants, identifies situations associated with an emotional reaction, and is taught to discriminate adaptive from unadaptive expressions of emotion.

(2) *Skill acquisition*, in which new ways of thinking about emotion-provoking situations are learnt, with emphasis on perceiving situations in a problem-solving way.

Included in stress inoculation may be self-admonition in the form of "internalized" verbalizations, such as "Don't personalize the issue", which appear to act as cues for the modification of emotionality. The value of this type of technique in the control of reaction to stress has yet to be properly evaluated but appears to be promising.

## Emotions

In the foregoing section the role of cognitions in the control of unwanted emotional reactivity was discussed. The emotions themselves, and direct control over them without the mediating influence of cognitions, deserve separate mention.

The range of unpleasant affective states and their power to influence behaviour have been well reviewed; they present numerous intriguing psychological problems (2). Schacter attempted to deal with some of the problems by means of theoretical formulations arguing broadly that a common physiological state underlies all emotional experience but the specific label given—sad, happy, etc.—depends on the cognitive attributes of the specific situation (12). Any way in which the situation were to be restructured cognitively might then lead to altered emotional experience—in the same way as is described above in relation to internalized speech. Now, however, the concern is to discuss the exercising of direct influence on the emotion itself.

Medication is important in this respect. For example, anxyolitic and antidepressant drugs can help to control negative emotional states. Specifically, central serotonin or norepinephrine deficiencies may be responsible for a depressed state and rectification by means of drugs is advisable. Problems result when such initiating conditions give rise to psychological response styles that themselves produce difficulties and a deterioration in the feeling state. The end product, clinically, is often extremely tangled but the chemical control of aberrant mood is frequently a first step towards unravelling the complexities involved (4). Nevertheless, there are behavioural methods that constitute direct approaches to the control of emotions and two deserve mention.

Common sense allows that fear can be overcome if it is possible to become accustomed to the "fear stimulus" gradually. A child, fearing the dark, seems to respond well to a graduated approach involving the progressive reduction of the amount of light in the bedroom over a period. Similarly, fear of water may be best overcome by encouraging the child first to play in shallow, warm water so that he learns to enjoy it, only gradually introducing more adventurous conditions as he begins to feel secure.

The behavioural application of this basic common-sense approach, called *systematic desensitization*, was developed as a therapeutic procedure by Wolpe (15). Briefly, the approach requires that the fear situation should be dealt with in stages and that, at each stage, there should be a deliberate attempt to produce a feeling state incompatible with the negative emotion. To secure success, considerable care must be exercised in planning the approach as well as in its execution (5). To a certain extent it seems that common anxieties can often be eliminated by using this approach, which can frequently be applied with profit in the imagination before proceeding to the real-life situation; a purely mental rehearsal of the steps or stages involved in facing a feared object or situation can help considerably.

An alternative direct approach to combating uncontrolled feelings has a quite different theoretical background and a completely different treatment procedure from that used in systematic desensitization. Learnt emotional reactions are characterized by avoidance reactions. The aim of treatment, it is argued, should be to re-create the original fear reaction without allowing avoidance behaviour. In practice, this means evoking intense emotional reactions and holding the individual in the situation that evokes them until they begin to diminish and eventually disappear. The name given to this type of approach, for which a reasonable amount of success has been claimed, is *implosive therapy* or *flooding* (14).

## Behaviours

It is difficult to appreciate that a deliberate change introduced into behaviour can exert an important influence on thoughts and feelings, yet social psychology is replete with examples of this in the context of attitude change. A clinical example was provided in a description of changes in sexual behaviour, achieved through conditioning, that affected the attitudes and feelings of those under treatment. Altering one component of a deteriorated marriage relationship—specific sexual behaviour—influences the way the couple think, feel, and treat each other in general (1).

An obvious way in which changes in behaviour can affect the experience of stress is, of course, through altering working practices. It is not uncommon—particularly among Type A individuals—to find a work schedule that allows very little free time, an early start and a late ending to the working day, little or no time set aside for meals (a sandwich eaten at the desk for example), few holiday breaks, lack of family life, no relaxation pursuits, and so on (8). To the investigator it remains a source of astonishment that the individual concerned seldom questions such an extraordinarily punishing schedule.

A change in this type of behaviour involves two important stages. Firstly, the faulty work practices must be meaningfully pointed out to the individual and the implications of continuing them very clearly explained; it is perfectly reasonable to spell out the consequences of a crippling work schedule if it continues to be pursued—e.g., the likelihood of serious physical disorder. Secondly a more reasonable approach to the work is planned with the person involved, and carefully recorded and monitored. However, personal needs must be taken into consideration; it may be a gross error of judgement to impose, except as a temporary measure, an ordinary working day on a captain of industry, a high-flying executive, or an entrepreneur. Although there is much more to be said about the details of this approach, at least this short description serves to illustrate an attempt to change behaviour by instruction and, in that way, to secure more stable emotional and cognitive states.

Some direct interventions at the level of overt behaviour are rather more formal in nature—e.g., training in social skills and assertiveness (9). Such interventions are necessary for individuals whose interpersonal or communicational skills are poorly developed and who, for those reasons, experience stress. A good example, and one often seen in the clinical setting, is the individual whose promotional prospects are affected because of this type of problem.

Typical components of the training are modelling, role playing, behavioural rehearsal, and feedback. Modelling means that the desired behaviour is exhibited by the trainer or therapist and thus provides a model for the participant to observe and emulate. Role playing means that particular ways of behaving are assigned to the participant, to act out in a manner both prescribed and modelled. Behavioural rehearsal means the continuous practice, or rehearsal, of a particular required way of behaviour, in which, it is hoped, new behaviours are consolidated. Feedback is information on performance given after practice so that performance may improve.

Understandably, such interventions involve numerous processes and mechanisms and are not simply deliberately contrived changes in behaviour; the emphasis is on explicit and overt aspects of behaviour and the way in which their specific modification leads to improvements in general. Some success is claimed for these methods, particularly when there is no formal psychiatric illness, with which such problems as lack of social skills and assertiveness are sometimes associated.

Further means of altering behaviour with a view to changing an individual's perceptions, cognitions, and feelings are often referred to as self-control. The objection most often raised to such means is that they would not be needed if self-control actually did exist. Yet this objection is not entirely valid, since there are very real possibilities for individuals to learn to take charge of their lives in ways that bear directly on the faulty behaviours they exhibit. The kind of action envisaged is often in accord with common practice—e.g., putting away a box of chocolates to avoid the temptation to eat one—and, in effect, means that the cues that prompt particular behaviour have been rearranged. Similarly, smoking may be curtailed by having a cigarette only under particular, infrequent, circumstances, or the impulse to overspend may be controlled by carrying only a small amount of money. It is important to be a good manager, so to speak, since the application of specific self-control procedures may require care and often a certain amount of resolution—e.g., making smoking unpleasant by placing quinine under the tongue before lighting a cigarette.

Self-control procedures seem deceptively simple and, unless well presented, can easily be dismissed as valueless (*10*). Yet, put into practice with a certain amount of energy and conscientiousness, they can have surprising and sometimes substantial effects on habits. Furthermore, when they prove successful, they tend to generate enthusiasm and motivation, which inspire greater effort and more change.

## Conclusions

Behavioural techniques have developed considerably in range and sophistication since Watson's first formulations.[1] Theoretically, views have gradually progressed from those represented by somewhat stark reflexology to at least a recognition of the complexity of human psychological functioning. Although it is something of a truism to say so, the need for scientific vigour in the realm of explaining and controlling human behaviour remains; for that reason alone the behavioural approach is a commendable one. Since behavioural theory and practice have been substantially concerned with abnormal and unpleasant states of psychological functioning, the implications for work stress are self-evident.

The categorization of psychological reactions into cognitions, emotions, and behaviours adopted in this chapter is arbitrary; in practice such distinctions are blurred. Indeed the thoughts,

---

[1] J. B. Watson, 1878–1958. A United States psychologist and leading exponent of behaviourism.

feelings, and actions of individuals in distress are invariably congruent and the relevance of which changed first or which is the more important may be little more than academic. Yet to make such distinctions can have practical value in the sense that, given the range and variety of techniques available, the choice of which ones to use may be influenced by the specific category most likely to gain advantage. While no doubt a package of techniques would be used, incorporating cognitive, emotional, and behavioural considerations, emphasis might be placed on one or the other as being particularly suited to the individual, the problem, or the particular circumstance.

The techniques are refreshingly direct, focusing as they do on the problem itself and not on assumed underlying conflicts or hidden mechanisms. Aberrant thought patterns are dealt with by erosion, or substitution; new behaviours are engineered to displace unsatisfactory or unadaptive ones; erratic and unreasonable emotional reactions are extinguished. The underlying idea, of course, is that change must involve the systematic and intensive application of techniques, since an exercise in learning and the consolidation of learning are involved.

# References

1 ASIRDAS, S. & BEECH, H. R. The behavioural treatment of sexual inadequacy. *Journal of psychosomatic research*, **19**: 345–353 (1975).

2 AVERILL, J. R. On the paucity of positive emotions. In: Blankstein, K. R. et al. *Assessment and modification of emotional behavior.* New York, Plenum Press, 1980.

3 BEECH, H. R. Learning: cause and cure. In: Cooper, C. L. & Payne, R. ed. *Stress at work.* Chichester, New York, Brisbane, and Toronto, Wiley, 1980, pp. 149–172.

4 BEECH, H. R. & VAUGHAN, M. *Behavioural treatment of obsessional states.* Chichester, New York, Brisbane, and Toronto, Wiley, 1978.

5 BEECH, H. R. ET AL. *A behavioural approach to the management of stress: a practical guide to techniques.* Chichester, New York, Brisbane, and Toronto, Wiley, 1982.

6 ELLIS, A. *How to live with a neurotic: at work or at home.* New York, Crown, 1974.

7 EYSENCK, H. J. *Behaviour therapy and the neuroses.* Oxford, Pergamon Press, 1960.

8 FRIEDMAN, M. & ROSENMAN, R. H. *Type A behavior and your heart.* New York, Knopf, 1974.

9 GALASSI, M. D. & GALASSI, J. P. *Assert yourself: how to be your own person.* New York, Human Sciences Press, 1976.

10 GOLDFRIED, M. R. & MERBAUM, M., ed. *Behavior change through self-control.* New York, Holt, Rinehart & Winston, 1973.

11 MEICHENBAUM, D. *Cognitive-behavior modification: an integrative approach.* New York, Plenum Press, 1977.

12 SCHACTER, S. *Emotion, obesity and crime.* New York, Academic Press, 1971.

13 SCHWARTZ, R. & GOTTMAN, J. Toward a task analysis of assertive behavior. *Journal of consulting and clinical psychology,* **44**: 910–920 (1976).

14 STAMPFL, T. G. & LEVIS, D. J. Essentials of implosive therapy: a learning-theory-based psychodynamic behavioral therapy. *Journal of abnormal psychology,* **72**: 496–503 (1967).

15 WOLPE, J. *The practice of behavior therapy,* 3rd edition. New York, Pergamon Press, 1982.

# Chapter 19

# The use of behavioural therapy in somatic stress reactions

H. Reginald Beech[1]

The term "stress" is employed in numerous ways, often producing obscurity rather than clarity. Many use it to describe an unpleasant feeling state, although it is apparent that such a state may come about in a variety of ways, including alterations in the biological processes of a natural kind, such as during the premenstrual period, or of a pathological kind, such as in schizophrenia. The individual usually thinks that life events are responsible for adverse feelings.

It is certainly impossible, within a meaningful definition of stress, to avoid reference to personal experience, but to understand it properly it is essential to determine those elements that conspire to produce it. There are four contributing components: basic vulnerability, coping skills, unpleasant life-events, and personality characteristics (1).

*Basic vulnerability* refers to a predisposition to react with exaggerated sensitivity to the less pleasant circumstances of life. It includes a disposition to worry unnecessarily about matters that are unpleasing but commonplace, to feel injured by comments made without rancour, and to experience changes in bodily state—a pounding heart or excessive sweating—in response to even mildly noxious events.

This quality of vulnerability probably forms a continuum along which all individuals may be arranged, some with a high and some with a low status on the variable. Those with a high status—or high vulnerability—tend to experience psychological discomfort with considerable readiness and intensity, often with relatively mild stimulation producing an adverse reaction; those low on the variable are capable of long exposure to strong and unpleasant stimulation before psychological distress or discomfort becomes apparent.

It is obvious that the experience of psychological discomfort termed "stress" by the individual is likely to be a function of vulnerability; the quality of feeling attaching to a "stressful" experience is directly related to the sensitivity of the individual involved.

*Coping skills* are the means by which feelings of psychological discomfort are either attenuated or set aside. They mainly consist of the strategies for the reduction of adverse feeling states that the individual has acquired, through instruction or

[1] Department of Psychiatry, The University Hospital of South Manchester, Manchester, England.

example, fortuitously or otherwise; they include skills in handling significant relationships and the ability to develop states antagonistic to anxiety, such as muscle relaxation. It is usual to include in this category certain attributes that could not properly be called skills, but rather coping aids, such as a supportive spouse and family.

*Unpleasant life-events* are what most individuals think of as the sources of stress, such as an unsatisfactory job, domestic strains, or other excessive environmental demands. The evidence available indicates that the organism requires an optimum amount of stimulation to retain psychological equilibrium; too little stimulation produces imbalance and stimulation that is too intense also results in abnormal functioning. When individuals complain of stress they are usually calling attention to the results of overstimulation; one or more intensely experienced event, or exposure to continuous overstimulation over a period, having produced psychological discomfort or distress. When such circumstances or events are implicated, attention must be given to reducing their influence on the individual—e.g., through deploying coping skills—or rearranging the life-style or circumstances of the individual to allow him to escape or avoid the noxious stimulation—e.g., by changing the patterns of domestic or work behaviour.

As far as *personality characteristics* are concerned, attention has been focused on work stress and the hypothesized Type A individual. Whether or not the elements alleged to form a cohesive entity are eventually established, the underlying assumption seems to be sufficiently cogent, namely, that certain individuals appear to adopt modes of behaviour that can lead to the imposition of strains and exposure to stress factors. This, in itself, may fail to produce the experience of stress should the individual be of robust temperament and/or possess the appropriate coping skills. However, it is not uncommon to find a well marked need to achieve allied to a sensitive disposition. Since success often requires toughness of character and the fruits of success are attended by additional burdens and responsibilities, such a combination often leads to stress reactions. The same may be said of other combinations of the four components, which may be favourable to the individual or may conspire to exacerbate the experience of stress[1]. The individual whose existing vulnerability is high, whose coping skills are minimal,

---

[1] The four components are formally assessed by a new scale devised primarily to examine their independence and to establish norms for various working samples (2).

and who has a "stress seeking" personality will be most at risk when life's vicissitudes accumulate, while someone with a stable temperament, possessing numerous coping abilities, and without a disposition to seek achievement or to compete, will experience stress only in the most adverse of circumstances.

In pointing to the complex contributing factors that serve to underwrite the stress experience, it becomes apparent that the nature of the associated emotional and behavioural expressions of discomfort will probably reflect that complexity. In the present chapter, however, the intention is to focus on the primarily physical/physiological forms of reaction.

## Physical/physiological reactions to stress

The importance of hormonal and chemical mediators in the body's response to stress has long been recognized but, in effect, there are three main bodily systems that appear to be particularly susceptible: the digestive, the muscular, and the cardiovascular (1). The range of symptoms attributable to a disturbance in any one of those systems is considerable, although certain common disorders, such as coronary heart disease, ulcers, and muscle tension, are the best known (see Table 10). The physical reactions of individuals to stressors vary and further research is necessary to explain why this is so. The answers will inevitably shed light on the mechanisms that mediate psychological factors to produce the symptoms of physical illness. The idiosyncracy of physical reaction may well reflect the involvement of many factors, including genetic predisposition, drug and alcohol abuse, poor nutrition, and lack of proper exercise, as well as climatic variables and, perhaps, the influence of psychosocial factors—e.g., adverse life events, work stress.

The most extensive investigation of the association between psychological factors and physical disorder was that involving the Type A behavioural pattern and coronary heart disease (7). The Western Collaborative Group Study, of over 3000 men aged 39–59 years, indicated that a substantial number of those developing coronary heart disease exhibited a particular pattern of behaviour, characterized by obsessive time consciousness, excessive ambition, extraversion, etc. (see Chapter 13). Such individuals, when subjected to social stressors, show significantly higher systolic blood pressure as well as increased heart rate and heart rate variability (7).

Table 10.  Possible physical and behavioural effects of stress

| System | Effect |
| --- | --- |
| Cardiovascular | Coronary heart disease<br>Hypertension |
| Dermatological | Eczema and other skin<br>  complaints |
| Gastrointestinal | Ulcer<br>Irritable bowel syndrome<br>Nausea and vomiting |
| Genitourinary | Frequent micturition<br>Impotence and orgasmic<br>  dysfunction |
| Immunological | Lowered disease resistance |
| Locomotor | Fatigue and lethargy<br>Raised incidence of<br>  inflammation of the connective tissue |
| Muscular | Tension headaches<br>Low back and chest pain |
| Respiratory | Asthma<br>Breathlessness<br>Hyperventilation |

The increase in coronary heart disease in some Western countries has focused interest on stress management and Type A behaviour. Concern with the Type A pattern of behaviour has led to programmes in which the behavioural characteristics are modified, largely through self-control and cognitive techniques. The aim is to make Type A individuals recognize the nature of their behaviour, develop alternative patterns, and ensure that the changes are constantly monitored so that they are preserved.

Understandably, a careful analysis of the individual's behaviour and its characteristics must precede any attempt at intervention. Central to such an analysis is the identification of the stimulus or range of stimuli that activate the individual's response. Since, quite frequently, there are subtle aspects of the situation to be observed, detailed record keeping by the psychologist, psychiatrist, or whoever is involved in the change process, is always advisable.

A clear-cut example of a relationship between physical abnormality and precipitating sources of stress is found in Mr A, who was investigated and treated by the author. Mr A was the managing director of a medium-sized company which was prospering well under his leadership. Nonetheless, he felt obliged to put his energies and talents to use in the company's interests to a far greater extent than perhaps he was capable of sustaining. In fact, he worked extremely long hours, spent substantial amounts of time travelling and away from home—which he disliked—and was unwilling to delegate responsibility if he could avoid it. He had personally built the company up to its successful level and felt, and behaved, as if it would fail without his continuing individual attention.

For some time Mr A suffered gastric discomfort of increasing severity until he could hardly ever eat without extreme pain and frequent vomiting. His weight fell dramatically and surgical intervention was advised. Unfortunately, the operation failed to provide the hoped-for relief and the surgeon suggested that psychological help might be considered.

The situation was more complex than can be detailed here, but Mr A's main problem appeared to be that his attitudes to work and the way he carried it out were faulty and unrealistic; these, therefore, formed the main targets for attention. In cases where there is a strong physical or behavioural reaction to very specific work or life stimuli it may be necessary to arrange to break the bond. In short, there existed for Mr A a strong conditioned association between eating almost any type of food, unless it was bland and finely minced, and pain and nausea. Accordingly, in addition to addressing the primary causes of the problem, a brief deconditioning programme was arranged which fortunately resulted in Mr A being able to eat a wide variety of food, including solids, in a reasonably normal manner.

In many ways this is a fairly typical case, since it illustrates clearly the complexities so often involved when an initial difficulty has become compounded by secondary problems by the time appropriate action is sought. Complex problems generally require complex solutions; in Mr A's case it was necessary not only to remove the primary cause of the problem—faulty work attitude and practice—but also to undo the secondary learning process that had occurred.

## Combining physical and behavioural approaches to work stress

While the psychological control of abnormal physical states is frequently indicated, there is undoubtedly considerable scope for applying physical forms of help. Medication of an appropriate kind can often be useful but there are important forms of intervention that deserve particular attention as examples of physical/behavioural control.

Almost invariably the stressors encountered in everyday life produce increased muscle tension which, in turn, tends to affect various bodily systems. Ordinarily, the muscles of the body are under a considerable degree of voluntary control but frequently control is impaired and, typically, the muscles remain in a state of contraction even though for them to do so can no longer be regarded as adaptive. The lasting state of contraction often leads to the experience of pain, particularly in sensitive areas such as the neck, shoulders, back, and head. Tension headache has become one of the most frequently observed consequences of stress experience and is often traceable to aberrant muscle activity.

The deliberate relaxation of high muscle tension, when it can be achieved, tends to produce not only a mental state of relative calm but also measurable and important physiological changes, such as a decreased heart rate and lowered blood pressure. When occupational stress results in continuously high levels of muscle tension, steps to counteract such a condition will have beneficial consequences, and may be helpful, for example, in the recovery of coronary patients. Ordinarily, such steps would involve preliminary instruction and a detailed programme of daily exercise, since the capacity to relax is a skill to be acquired through practice (4).

The exercises involve the progressive contraction and relaxation of the muscles in the major groups, enabling the individual to acquire control through deliberately increasing muscle tension and then creating the contrary state. Typically, several weeks of daily practice are needed to acquire a viable level of skill. However, this may not suffice in some cases and recourse to a more sophisticated method is required, such as biofeedback, a technique of particular interest in the context of abnormalities in physical functioning.

While the separateness of the psychological and physical areas of functioning may be a matter for doubt, a case could be made for claiming that some physical processes are outside the individual's voluntary control. Indeed, it may be argued that the involuntary nature of some physical mechanisms is essential, since their effective functioning must continue without awareness and constant conscious monitoring on the part of the individual. Nevertheless, a measure of conscious control is possible even over the most involuntary of bodily systems—e.g., by holding the breath or breathing more rapidly it is quite easy to alter the heart rate. Similarly, by changing mental content— e.g., entertaining a frightening thought—changes can be effected in heart rate, respiration, skin resistance, and degree of muscle tension.

It can, therefore, be observed that the physical and psychological processes are interdependent to a great extent. Biofeedback investigations are, in part, designed to explore the possibility of altering involuntary physical processes in a systematic way.

Essentially the feedback provided takes the form of information of a kind not usually available to the individual about the physical function involved. For example, the feedback given on heart rate could take the form of a digital display, a moving needle, or a variable tone, reflecting, and sensitive to, any changes. The task of the individual would be to try, by one means or another, to consistently affect the digital display, the needle, or the tone. In achieving this, the physical function is, of course, itself being affected.

## Headache and biofeedback

Head pains are a common stress reaction and, if frequently experienced, may severely affect work performance. The traditional categorization of this locus of pain is into tension headache—the cause of which is assumed to be excessive muscle tension—and vascular (migraine) headache—the cause of which is assumed to be irregularity in the blood flow. It has been suggested, however, that a continuum is involved rather than discrete categories of head pain, and biofeedback treatment has been given to both tension headache and migraine sufferers (9).

An 80% success rate was reported for tension headache cases given electromyogram biofeedback treatment, which was directed at reducing tension in the muscles thought to be culpable in producing the sensation of pain (3). A high success rate was also reported in the biofeedback treatment of migraine patients who had remained unresponsive to the range of medication ordinarily offered. The investigators based their approach on the assumption that the provision of feedback on body temperature would be advantageous and raising the hand temperature—frequently low during migraine attacks—particularly efficacious (10). The patients showed a ready capacity to increase hand temperature after a period of training, and this newly acquired skill appeared to be associated with the control of migraine headaches.

Mr B was a 45-year-old restaurant manager admitted to a neurological unit for investigation after having suffered constant vascular headache for 4 months. The severity and persistence of the pain had led him to consider suicide, particularly as he was unable to secure relief through medication. He was given an appointment for 2 weeks after his discharge from the neurological unit but arrived at the psychology department on the day of discharge in desperation, asking for immediate help. He was given temperature biofeedback treatment and even during his first days of training was able to raise his hand temperature by a remarkable 7°C (typically 1–2°C can be achieved). His accomplishment brought him instant relief from pain and he readily acquired the capacity to eliminate further attacks, thus controlling his suffering, without the need of biofeedback apparatus. Some 7 years later he is still able to work, free from head pain, in spite of numerous life stresses. He attributes this to the control acquired during his biofeedback treatment.

The picture, however, is certainly not always so simple and clear cut as that described above. While many individuals show the generally assumed relationship between exposure to stress and a migraine reaction, others, paradoxically, show evidence of "weekend migraine". The latter individuals appear to function well in the presence of work stressors but suffer severely when the stressors are removed, during a time of relaxation when they might expect to be free from problems. Yet biofeedback treatment for head pain has proved very successful, whether the approach used has involved the control of muscle tension or the control of body temperature. Of particular importance is the fact that this has been without the adverse side-effects on work often associated with drugs.

## Biofeedback and hypertension

High blood pressure is one of the most common physical disturbances in Western societies, the most frequently encountered form being essential hypertension, a condition for which no obvious organic pathology appears to be responsible. High blood pressure places the individual at considerable physical risk and any decrease in level tends to be associated with increased life expectancy. It is well known that stress increases the blood pressure of many individuals; repeated exposure to stressors appears to lead to permanently elevated blood pressure (8). While medication is the most frequently applied treatment for the condition, drugs are not always appropriate for one reason or another, and an alternative approach may be indicated.

Early studies on the biofeedback treatment of patients with blood pressure indicated that it could be helpful, provided it was intensive. Relaxation training has also been demonstrated to be useful (5). In a study of executives and skilled workers who had been identified as being hypertensive during as annual company medical examination the feedback of skin temperature, together with relaxation training, was employed in an occupational health setting (12). It was found that most subjects could learn to produce significant reductions in blood pressure. This study is of particular interest since treatment occupied only 20 minutes a week and, as it was given at the place of work, was minimally disruptive to the working schedule.

## Myocardial infarction

It has long been recognized that psychological factors contribute to the pathogenesis and onset of heart attack (11). In particular, anxiety and depression have been associated with heart attack and emotional factors have been involved in cardiovascular complications. A diminished risk of coronary heart disease and a possible improvement of state following myocardial infarction could, therefore, accompany the treatment of emotional states.

In a mixed approach to the rehabilitation of myocardial infarction patients a programme that included both counselling and relaxation training proved to be beneficial (6). The results of treating patients newly discharged from a coronary care unit indicated that improvements were mainly due to acquiring skill in muscle relaxation. Although the data available covered only a 10-week period of observation it was suggested that, since high levels of autonomic arousal may lead to increased cardiac contractile force, relaxation could be beneficial by reducing the former; hence the duration of survival following myocardial infarction would be increased.

## Conclusions

It could be argued that the methods of control listed in this chapter are hardly likely to be effective for work stress, since they emphasize the outcome rather than the cause. Surely, it could be said, if the culpable working conditions, or a variable associated with them, still apply, unfavourable reactions of a physical or psychological character could still result; the only truly satisfactory means of eliminating a stress reaction is to eliminate the stressor.

There is a degree of validity in such an argument; when an individual has a punishing work schedule imposed upon him, or one that is self-imposed, for example, a change in that schedule would be an essential preliminary to securing relief from stress. But the experience of stress is most typically a product of at least four major influences, only one of which is the nature and extent of the stressor itself. Furthermore, it is evident that much of what is described as a stress reaction is a kind of residual habit that has persisted and become, as it were, functionally autonomous—the unadaptive reaction continuing long after the actual stimulus to stress has been removed.

It is not difficult to appreciate how the latter effect may come about in quite common and simple circumstances. For example, the loss of work through redundancy, may, and often does, produce thoughts, attitudes, and motor behaviours that become fixed and affect future job prospects, thus further increasing general misery and ineffectuality. Physical and behavioural reactivity, at one time the response to stressors no longer active, may well persist as a functionally autonomous habit and can be dealt with on that basis. A common example of this is muscle tension and its attendant aches and pains which, when dealt with in a direct manner, can be eliminated without fear that a new form of abnormality will appear. It seems quite likely that essential hypertension may fall into the same category.

Most often, however, occupational stress involves maladaptive reactions in a set of interrelated areas of functioning. To achieve the greatest effectiveness, treatment should be directed to each of the affected areas—cognitive, emotional, motor behavioural, and physical—within a framework that takes account of the temperament and personality of the individual, his level of coping skills, his behavioural patterns, and the actual stresses and strains suffered.

Not surprisingly, the behavioural approach to the understanding and treatment of occupational stress reactions must be regarded with certain reservations. Perhaps its most salient shortcoming is its failure to accommodate permanent, or occasional and episodic, alterations of internal state within the one theoretical and therapeutic system. There are individuals who, because of a serious and basic vulnerability, are unable to usefully occupy positions in which high stresses and strains are inherent. Some positions very clearly require considerable robustness of character and as much importance needs to be attached to this as to any other personal quality. This is often overlooked, however; training, aptitude, personal ambition, and other similar qualities are given appropriate and careful weighting but little or no attempt is made to assess vulnerability.

In essence, the ability to assess relevant factors and to draw conclusions concerning the areas and extent of intervention in the realm of work stress are skills much dependent on experience. A behavioural approach offers clarity and purposiveness and for those reasons is commendable. It also requires experience. In behavioural terms, just as for any other theoretical framework, an accurate statement of the problem is an essential prerequisite to finding a solution.

## References

1  BEECH, H. R. ET AL. *A behavioural approach to the management of stress: a practical guide to techniques.* Chichester, New York, Brisbane, and Toronto, Wiley, 1982.

2  BEECH, H. R. & RADELAAR, H. O. F. *4-component stress scale.* Department of Psychiatry, University of Manchester, 1982 (unpublished).

3  BUDZYNSKI, T. H. ET AL. EMG biofeedback and tension headache: a controlled outcome study. *Psychosomatic medicine,* **35**: 485–496 (1973).

4  BURNS, L. E. Relaxation in the management of stress. In: Marshall, J. & Cooper, C. L., ed. *Coping with stress at work.* London, Gower Press, 1981, pp. 95–110.

5  COOPER, C. L. *The stress check.* Englewood Cliffs, NJ, Prentice-Hall, 1980.

6  FIELDING, R. A note on behavioural treatment in the rehabilitation of myocardial infarction patients. *British journal of social and clinical psychology,* **19**: 157–161 (1980).

7  FRIEDMAN, M. & ROSENMAN, R. H. *Type A behavior and your heart.* New York, Knopf, 1974.

8  MELHUISH, A. *Executive health.* London, Business Books, 1978.

9  PHILIPS, C. Tension headache. *Theoretical problems,* **16**: 249–262 (1978).

10  READING, C. & MOHR, P. D. Biofeedback control of migraine: a pilot study. *British journal of social and clinical psychology,* **15**: 429–433, (1976).

11  ROSENMAN, R. H. & FRIEDMAN, M. Behavior patterns, blood lipids, and coronary heart disease. *Journal of the American Medical Association,* **184**: 934–938 (1963).

12  SHEFFIELD, B. F. Biofeedback and stress. In: Beech, H. R. et al. *A behavioural approach to the management of stress: a practical guide to techniques.* Chichester, New York, Brisbane, and Toronto, Wiley, 1982, pp. 97–114.

# Chapter 20

# The role of the occupational health professional at the place of work

Alan A. McLean[1]

## Introduction

In his introduction to a report on research on stress and human health, Dr David Hamburg, President, from 1975 to 1980, of the Institute of Medicine, United States National Academy of Sciences, said that during his term of office no aspect of health and disease had elicited more interest among leaders of the Government of the United States of America. Senators, congressmen, cabinet officers, and other leaders had repeatedly expressed concern about the possible effects of very difficult life experiences on health.

What can employers do to reduce stress on the job and to implement programmes for the prevention and treatment of stress reactions and mental health problems? What is the rationale for introducing an occupational mental health programme? How can psychiatric disorder in the world of work be prevented and/or treated, in both the broad and the narrow sense? Such questions can be considered under two headings. One relates to the organizational practices that may be helpful in reducing levels of stress among employees; why and how organizations should put into effect activities to reduce stress reactions and improve the mental health of their workers. The other relates to specific activities for the early detection of emotional problems and for appropriate intervention and treatment; the kinds that are helpful in aiding employees as they cope with stress reactions and other psychiatric disorders.

## Rationale for introducing an occupational mental health programme

Sound occupational health activities reflect the concern of managers for the welfare of employees, particularly if they are seen as part of an overall attitude of concern and consideration for each employee's legitimate needs while at work. In a more concrete way, most employers have little desire for the working environment to allow exposure to toxic substances. The more sophisticated employer is equally concerned about the emotional climate, and will take steps to forestall the development of one that is receptive to the seeds of conflict and stress. The

[1] International Business Machines Corporation, New York, United States of America.

components of an occupational health programme include sound management practice—i.e., firm, fair, and consistent management techniques; a benefit programme with sufficient coverage to allow prompt and reasonably complete preventive and therapeutic health care; and, for the larger employer, within-company employee assistance and counselling programmes at various levels of professional competence.

In the USA, fewer stress-related claims on workers' compensation schemes mean less cost to the employer. This has a very important economic effect, since workers' compensation costs can make up a significant portion of payroll expenses. In addition, other legal liabilities to the employer are important reasons to support any activity that can reduce employee disability. Another reason for an employer to support a programme related to occupational mental health is that the term "stress" has become more popular. For many, "job stress" or "work stress" are far more acceptable terms than "mental health". The experience of stress does not have the stigma of mental illness; in order to fight against it and its effects an employer can provide a range of within-company activities and utilize community support systems that might not be so acceptable if spoken of as "psychological" or "psychiatric".

Of significant concern to many employers is the rise in health care costs under various benefit schemes for which they pay. Benefits covering the treatment of psychiatric disorders seem the least concrete and manageable, yet, for many, represent the most rapidly growing cost. For example, for a company in the USA with a total covered population of 650 000—208 000 employees and 442 000 dependants—the 1980 mental health benefit costs amounted to US$ 26.6 million—US$ 14.9 million for inpatient care and US$ 11.7 million for outpatient care. A most startling fact is that in 1967 inpatient psychiatric care costs represented 6.4% of the total hospital benefits paid out, whereas in 1980 they represented 15.4%. In other words, in 13 years inpatient psychiatric care costs nearly tripled as a percentage of the total hospital costs paid out. In 1980 the average amount of a claim for the cost of hospitalization for a psychiatric patient was double that for the next diagnostic category, cardiac disorder. This serves to illustrate an additional reason why an employer should be interested in having an occupational mental health programme. If the development of even a few cases of disabling mental illness can be forestalled, potential advantages will accrue in terms of corporate costs. In any event, the occupational health professional, including the mental health professional, is becoming increasingly involved in working with employers and their insurance carriers to assist in workers' compensation and benefit cost management problems.

There are, therefore, at least four reasons why employers should introduce occupational mental health programmes: enlightened self interest, their workers' compensation liabilities, the increasing cost to benefit schemes of psychiatric care, and the recognition of stress at work as a part of life.

## Programmes for the care of emotional, behavioural, and psychosomatic disturbances

Employee mental health programmes are hardly a recent phenomenon. In the USA some were well established in the 1920s. Many came into being during the Second World War and dozens more, based on essentially the same model, were started in the 1950s and 1960s. For the most part they are staffed by psychiatrists associated with occupational health programmes working either full or part time. There is a rich and ample literature documenting their existence and detailing their activities (2).

Operationally, the early "industrial psychiatry" programmes placed emphasis on individual employee-patient diagnostic evaluation, and some therapy. The psychiatrists' main tasks were to provide consultation and advice to other physicians in medical departments, to employees, and very often to management. The advice, counsel, and therapy they gave traditionally encompassed the care of emotional, behavioural, and psychosomatic disturbances in the employees. Some became involved in applied research, studying populations in an effort to determine the processes in the working environment through which anxiety was unnecessarily activated or created or which gave rise to stress in other ways. Many were involved in education and in management development activities.

Many psychiatric consultants currently working in industry continue to engage successfully in the same functions. Among the advantages derived from such functions is the fact that psychiatric disorder can be detected early in its disease process and the initiation of treatment to forestall more disabling illness encouraged. The same functions pertain to the organization itself, for many psychiatrists regard the organization as the patient as much as the individual employee. As could be expected, their areas of interest, concern, and activity have expanded in the last decade or so to include such areas as social support systems, cardiovascular risk factors, mental health benefits, organizational values, and the changing meaning of work. Accidents and the abuse of drugs and alcohol have been of continuous concern. Since the occupational psychiatrist is liable to encounter a wide variety of illnesses, he must remain professionally competent and alert. He has also been one of those participating in the increasing interest shown in stress at work *per se*.

Activities grew from the above-mentioned historical base, diversified, and prospered in many areas. Non-medical, and often non-clinical, workers from many disciplines selected various aspects of what was a fairly coherent field—occupational mental health. Psychiatrists, clinical psychologists, and occupational physicians have all participated in organizational mental health care, but so have many who do not have such a high level of training and expertise. Thus some assistance programmes that offer support for employee mental health are staffed by psychiatric social workers and some are quite good. Many deal exclusively with the employee who is a problem drinker and are, for the most part, staffed by counsellors on alcohol-related problems with little or no clinical training and often no professional background.

Occupational physicians themselves are becoming more sophisticated in dealing with psychiatric issues and many in human resource departments have gained a greater understanding of the psychodynamics of human behaviour. A specialized field in organizational development has emerged; and efforts to improve the quality of working life are accelerating, with greater employee participation being urged, especially in decision-making as it relates to the specific tasks and projects in which they are involved. Occupational mental health is coming to be overshadowed by new and increasingly popular concepts— exemplified by the terms "work stress" and "burnout". There are now thousands of stress management "experts" in the USA, seemingly more numerous than the 4000 occupational physicians, and the several hundred occupational psychiatrists with more traditional expertise and training.

Various types of stress management programmes have come into vogue. They may be independent, or part of an occupational health department. Regardless of where they are located in the organization, however, their purpose is to teach employees the techniques that will enable them to cope with, or prevent, anxiety and tension. The techniques involved include biofeedback, muscle relaxation, and meditation. Related to them are the many physical fitness programmes also sponsored by employers, some concerned only with cardiovascular fitness, others with a broader range of physical wellbeing. Although the goal is to enhance feelings of physical wellbeing, there is evidence that many such activities also enhance mental wellbeing.

## Stress and absenteeism

In a study of stress and its effect on absenteeism, the effect on absence through illness of health evaluation interviews after medical examinations was measured in a group of 500 corporate employees reporting varying amounts of stress (3). The interviews were designed as assessment tools to determine the effects of stressors at the place of work on the individual and to provide health counselling. As part of their medical histories the employees were asked to rate their experiences with 15 possible occupational stressors on a 5-point scale. An average was then computed for each individual which gave an individual job stress score—a numerical coefficient of stress at work. In a subsequent 20-minute interview stress-related symptoms were assessed, as was each individual's coping ability. The employees were then referred to sources of personal assistance. Educational materials were also provided. The referrals were made in an effort to assist employees by providing resources to promote health maintenance, decrease their medical claim costs, and utilize available within-company services. The employees were referred for counselling, advice on nutrition, exercise, and various biofeedback and relaxation activities. They were also sent to cease-smoking programmes and a variety of medical specialists.

The illness absenteeism rate was monitored 6 months before and 6 months after the interview. A control group whose members had not been interviewed but had been medically examined was matched by sex, job classification, and job stress score and was compared with the experimental group. Absence through illness in the group that had been interviewed and referred for assistance had decreased significantly; by 50% in the 6 months after the interview. The control group, on the other hand, showed an increase in absence through illness in the same period.

Thus it was demonstrated that, within the framework of an existing occupational health programme, assessment by a mental health professional, together with counselling and referral, is an extremely effective device in reducing short-term disability leading to absence through sickness.

## Alcoholism

Most mental disorders occurring in the working environment present a fairly striaghtforward challenge to the occupational health professional. With accurate diagnosis and a supportive management, appropriate restrictions can be introduced. A job in keeping with the temporary or permanent disability can be assigned. Intelligent advice can be given to the employee's supervisor. This, of course, implies a thorough knowledge on the part of the occupational health professional of the work organization, the specific job, and the existence of possible alternative assignments.

There is one disorder that presents a unique challenge and an opportunity for the staff of an organization's medical department to be effective. This is the disorder found in an individual whose consumption of alcohol has reached the point where it interferes with job performance. Many programmes exist to help such individuals. One, that has been in place for more than 25 years and that maintains a superb record, can serve as a model applicable to most work settings large enough to support a medical department. A report was published on 1154 employees referred to the programme (1). Of the 1154, 75% were referred by their supervisors, 15% referred themselves, and 10% were referred by medical department staff. Data were available for the 5 years prior to entry into the programme and the 5 years after. Five years after entering the programme 75% were considered to be either rehabilitated or improved. Five years before entering the programme only 10% had received a positive appraisal; 5 years after 66% were considered to be good employees. Five years before entering the programme 1080 of the 1154 employees investigated were sickness disability cases;[1] five years after the number had decreased to 522. Accidents, both at and away from work, also decreased by half. Roughly translating the absence data into cost–benefit figures, absence was reduced by 31 806 days and the company saved US$ 1 272 240.

The staff of the programme consists of two alcoholism counsellors and an occupational physician, who work in close

[1] Defined as more than 7 days of reported absence through illness in a year.

cooperation with a psychiatric consultant who sees about a third of the employee-patients. It is a model for industry not only in the way its experiences are recorded but also in its extraordinarily wide acceptance by employees, unions, and management.

## The role of the occupational health professional

Clinical contact with employee-patients in an occupational health department is an interesting part of professional practice. Usually the patients there are seen at a much earlier stage of illness than patients at a psychiatric clinic. Because many are less seriously disturbed than patients at a mental health facility, treatment can often be relatively brief. In an occupational health programme, prolonged care is not practical but a vast amount can be done in a short time for a responsive disturbed worker.

As part of his clinical activities, the psychiatrist is often asked to assist in evaluating the mental status of an applicant for employment during the recruitment medical examination if the applicant has presented a history of psychiatric disorder or the examining physician has questioned his ability to adapt to the assignment for which he is being considered. This sort of evaluation is not carried out with a view to excluding the applicant from employment but rather to finding the best person–job fit. The sort of evidence he looks for is whether there is a likelihood that the applicant will find the work so stressful that it will produce disability; whether there is a match between the applicant's skills and the demands of the job; whether the applicant is so handicapped that unsuccesful performance might be anticipated.

The occupational health staff may be called on when a patient is due to return to work after having been hospitalized for a mental illness, to help to ascertain how the working environment can be most supportive in rehabilitation. Employers differ widely in their degrees of willingness to be helpful in this regard. Many, however, are flexible enough and have the motivation to modify jobs sharply, and at times to create work for such employees. In many countries there are laws and regulations that encourage employers to make special provisions for the handicapped.

Perhaps the most important clinical role of the occupational health professional is not in direct patient contact; it is in working behind the scenes with others, both in the medical department and in the management hierarchy, that his expertise can be of greatest benefit. As the psychiatrist and the sophisticated occupational physician teach others to understand and to cope with deviant behaviour, the knowledge spreads until it becomes manifold. How to interpret the dynamics of malfunctioning groups of workers and to facilitate their effectiveness can come to be understood in the same way.

Promotion, demotion, job transfer, and the need to adapt to new technology are ways of life at work and are often stressful and difficult to accept. A function of the staff in an occupational health programme is individual consultation with those undergoing such changes. Supportive counselling is often particularly valuable at times of career change. Other functions include the initiation of applied research activities to determine the characteristics in the working environment that adversely influence, or support, healthy behaviour. This may involve the gathering of epidemiological data on the distribution of mental disorders among employees; and the study of various subgroups within an enterprise in efforts to ascertain the factors that produce unnecessarily high levels of anxiety. Finally, there are the many nonclinical functions that directly or indirectly affect the mental health of the employees of an organization.

## The relationship of the occupational health professional to management

The psychiatrist and the occupational physician have identities in the work setting as experts in deviant human behaviour. Their roles are clearly recognized but they themselves and the management of the organizations that employ their services may interpret them differently. It is, therefore, necessary at the outset of the occupational health professional's relationship with the organization to ensure that his responsibilities and the boundaries of his functions are clearly defined. That his role is consultative should be made clear. He can certainly evaluate and advise but it should be recognized that administrative responsibility and the decision-making, in sofar as it involves non-medical policy, rests with the management. This may be seriously frustrating for someone who, in other settings, is accustomed to being regarded as the ultimate authority. For example, an employee with a schizophrenic illness may be urged by the management to resign if he does not wish to be discharged; the physician may have strongly recommended that he be retained and efforts be made to rehabilitate him.

In both the public and the private sectors, the mission of the organization is to grow and to survive. Most are not in business to provide programmes devoted to employee health or, indeed, working environments conducive to job satisfaction. Since most organizations are not principally concerned with health care they often do not encourage efforts to control emotional, behavioural, or psychosomatic disturbances on the job. That does not mean, however, that appropriate care cannot be arranged elsewhere and that intensive systems of followup are not appropriate to the organization's professional occupational health personnel.

## References

1 HILKER, R. Alcohol abuse: a model program of prevention. In: McLean, A. et al., ed. *Reducing occupational stress.* Washington, DC, United States Government Printing Office, 1978 (DHEW Publication No. (NIOSH) 78–140).

2 McLEAN, A. Occupational psychiatry. In: Kaplan, H. I. et al., ed. *Comprehensive textbook of psychiatry,* 3rd edition. Baltimore, Williams & Wilkins, 1980, Chapter 47.

3 SEAMONDS, B. C. Stress factors and their effect on absenteeism in a corporate employee group. *Journal of occupational medicine,* **24**: 393–397 (1982).

# Part Six

# Future research on psychosocial factors at work

# Chapter 21

# Future research
Lennart Levi[1]

## Areas of research

The basic assumption for research on psychosocial factors in occupational health is that they can precipitate or counteract ill health, influence wellbeing, and modify the outcome of health measures. Based on that assumption, a framework was suggested in Chapter 2 (Fig. 1) for determining important areas of research, involving: (1) the social structures and processes at and outside the place of work, viewed objectively and as perceived by the employee—i.e., potentially pathogenic environmental events, or stressors, their measurement and their interrelationship; (2) the vulnerabilities and other characteristics of the employee—i.e., the employee's propensity to react pathogenetically to such psychosocial stimuli and the genetic and environmental determinants involved; (3) the pathogenic mechanisms in the employee's reactions—i.e., the cognitive, emotional, behavioural, and physiological mechanisms of pathogenic significance; (4 + 5) the consequences, in terms of health and wellbeing; and (6) the interacting and modifying influences—i.e., the specific factors that modify interaction among (1)–(5).

Most available research findings demonstrate either that (1) and (4 + 5) are associated, or that (1) can produce (3) in all individuals or just in those with specific vulnerabilities (2). There is little or no information on how (3) leads to (4 + 5), on the buffering or facilitating significance of interacting variables (6), or on the—probably non-linear—interactions in the entire, cybernetic, man/working environment ecosystem [(1)–(2)–(3)–(4 + 5)–(1), modified by (6)](1).

## Hypothesis testing and evaluation of health action

In response to the greater than ever demand, considerable numbers of studies on occupational health and its environmental determinants have been carried out. Advances have been made at the molecular, cellular, and organ level but very little at the group, industry, or community level. Referring to this gap, it has been pointed out that once attention moves from the laboratory to the community reports on hypothesis testing studies are

[1] WHO Psychosocial Centre, Laboratory for Clinical Stress Research, Karolinska Institute, Stockholm, Sweden.

hardly ever found (2). Once it is wished to look beyond clinical trials, it is rare to find an evaluation of health action. Innumerable studies have shown that there are associations between psychosocial factors at work and health, have speculated on ideas for health or social action, or have put forward hypotheses in relation to the community spread and control of disease. In rare, important exceptions a hypothesis has been tested or action has been evaluated; showing that it is possible to carry out such a type of study (2, 6). Reasons given for not doing so are that it is unethical, technically impossible, or too expensive, or that it takes too long. Whilst there is an element of truth in all four objections, the first three can often be overcome, and something can be done to reduce the effects of the fourth (2, 5). Once such research is considered, however, it becomes unethical, and probably more costly, to impose an environmental or health action of still-to-be-proved value, possible danger, and high cost without evaluation, or to accept the hypothesis without prior test (2). Therefore, the first need is to determine from a high risk point of view the psychosocial factors in the working environment, the groups of workers exposed to such factors, and the reactions that can be harmful to health and wellbeing to which such exposure gives rise—i.e., cognitive, emotional, behavioural, and physiological. With such basic knowledge, measures that may be presumed to prevent disease and promote health can be proposed, which, at a third stage, can be evaluated in an interdisciplinary, experimental model study.

## Approaches

To obtain the necessary data, the approach to a research programme can often comprise the three following consecutive steps:

—determination of the environmental and health problems, using survey techniques and morbidity data;
—an intensive longitudinal, multidisciplinary, controlled field study of high-risk situations and high-risk groups and the ways in which they intersect;
—controlled intervention, which includes laboratory experiments, and evaluation of therapeutic and/or preventive interventions in real-life settings (evaluation of health action)—e.g., natural or field experiments, utilizing results from the other two steps.

A second, complementary, approach is to test key hypotheses, leading to an increased understanding of the psychosocial factors and health concept in general (3).

Both these approaches are important; both take time and consume resources. The first—leading to an evaluation of health action—should be used whenever there is a need to increase effectiveness, safety, and efficiency—i.e., in relation to nearly all community health action. The second—increasing general understanding—can sometimes be used at the same time and with almost the same resources as the first, at almost no additional cost, since the two are complementary. The ideal strategy would therefore be to select situations where the two can be combined efficiently, making it possible to engage in basic research and simultaneously to address the enhancement of occupational health care, the prevention of occupational ill health, and increased wellbeing, using not many more resources for the four combined than if each were to be addressed on its own. During a combined evaluative and hypothesis testing study it should also be possible, for little additional cost, to obtain information on the quantitative relationships between factors or elements in the working environment–stress–health system thought to be relevant to the outcome.

## Interventions to be evaluated

In other chapters of this book some of the existing ideas for improving the ways in which people can adapt their working environment to their abilities and needs are discussed. Most of the ideas have yet to be tested adequately, even on a small scale, particularly in ways that incorporate the interdisciplinary evaluation needed to determine the positive and negative consequences that may follow in terms of individual health and social demands. Among the interventions that, evidence suggests, could increase in effectiveness through additional research are the following (6):

— increasing the worker's control of the working arrangements;
— providing mechanisms for the worker to participate in decision-making on the organization of his work;
— avoiding imposing monotonous, machine-paced, and short but frequent tasks on the worker;
— optimizing automation;

—helping the worker to view his specific task in relation to the total product;

—avoiding quantitative work overload or underload;

—facilitating communication and support systems among workers and others.

As more information on these and other interventions becomes available from both developed and developing countries, it should be possible to begin to explore the difficult, but critical, issues related to the balance between social needs for productivity and individual requirements for good physical and mental health. Already available evidence suggests that, in some instances, an adequate understanding of the stressful aspects of a work setting may make it possible to improve both.

## Areas benefiting from research

Research will be of great benefit in making clear distinctions among stressful social structures and processes, reactions to such stressors, the consequences of such reactions, and the mediators that modify the flow of events. Information is also needed on the critical determinants that make some events and conditions stressful, on the effects of the resulting stress on a broad range of possible pathogenic mechanisms, and on the resulting health consequences. Information is similarly needed on the components of salutogenic processes and how they interact (1 and Chapter 15).

In the past, most studies were unifactorial, in the sense that they focused on a single characteristic of the work situation— e.g., machine-paced work—or of the worker—e.g., Type A behaviour—relating it to one possibly pathogenic reaction—e.g., catecholamine excretion—and/or to morbidity in a specific disease—e.g., myocardial infarction. Since the mid-1970s they have become increasingly multifactorial and even interdisciplinary. The next step ought, therefore, to be to apply the non-linear, interactional systems approach, advocated in this chapter, to the entire sequence of events, starting with the work situation and ending with the advantages of having healthy and contented workers and including intervening and interacting variables and feedback loops.

## Composition of a research programme

Research is not an abstraction; it is a tool that can help to provide the answers to questions about the cause, prevention, and treatment of ill health and human suffering (7). Accordingly, a research programme—not just an international or national programme but a programme in each community, on each shop floor, and in each office—could be designed for implementation in the following manner (5):

(1) Determination of the type and extent of the problems present—e.g., mental and psychosomatic disorder, absenteeism, alcohol abuse, labour turnover, dissatisfaction, or social unrest.

(2) Assessment of the psychosocial and physical environmental correlates of the various problems, in the work setting and outside it.

(3) Consideration, by community leaders, managers, members of labour unions, occupational health and safety workers, and other authorities, with the cooperation of the workers concerned, of the environmental influences likely to be of greatest causal importance, whether they are accessible to change, and whether, if changes are proposed, they will be feasible and acceptable to all concerned.

(4) Introduction of change in the working environment on a small and experimental scale, the evaluation of the resulting benefits and side-effects, and, on the basis of the evaluation, a decision on which of the changes could be introduced on a wider scale.

(5) Continuous monitoring and evaluation of the effect of the changes introduced on a wider scale, and modification as necessary.

To be efficient, the programme must include a means of providing feedback information on the results and must be carried out with the full participation and understanding of all concerned.

Such a strategy would provide solid answers to six key questions (7):

(1) What groups of people are at a high risk of developing mental illness or emotional disorder?

(2) What factors contribute to the risk and what is the relative importance of each of those factors?

(3) Can the most significant of the risk factors be effectively reduced or eliminated?

(4) Does eliminating the most significant of the risk factors effectively lower the rate of emotional disorder or mental illness?

(5) If it does, are the costs of intervention justified by the benefits obtained?

(6) Is the programme responsive to the principles governing both the rights of individuals and the rights of society?

## Targets

The type of research programme described above should lead to the following developments (4):

(1) General recognition that psychosocial factors must be taken into consideration in the field of occupational health.

(2) Recognition by managers of the fact that psychosocial factors exist. This will result in the humanization of working life and an appreciation of the fact that the worker's feelings of participation, belonging, job satisfaction, etc., are important elements of the work setting.

(3) A large number of people working in occupational health trained to recognize and influence the psychosocial risk factors in the working environment, and the various forms of health problems related to them. This newly acquired capacity is already being put to use in both developed and developing countries in occupational health care practice and in carrying out research.

(4) Continuous improvement of the psychosocial working environment and evaluation of the effects of improvement.

(5) Progress in creating and enhancing legislation in relation to psychosocial factors at work.

(6) The greater availability of literature, other information, and training materials on the recognition and modification of risk factors and health effects.

(7) The preparation for general use of specific methodology kits that include, for example, questionnaires.

(8) The organization of international congresses and workshops for the dynamic exchange of experience and knowledge.

## References

1 ELLIOTT, G. R. & EISDORFER, C., ed. Stress and human health: analysis of implications of research. New York, Springer, 1982.

2 KAGAN, A. R. A community research strategy applicable to psychosocial factors and health. In: Levi, L., ed. Society, stress and disease: working life. Oxford, New York, and Toronto, Oxford University Press, 1981, Vol. 4, pp. 339–342.

3 KAGAN, A. R. & LEVI, L. Health and environment—psychosocial stimuli: a review. In: Levi, L., ed. Society, stress and disease: childhood and adolescence. London, New York, and Toronto, Oxford University Press, 1975, Vol. 2, pp. 241–260.

4 KALIMO, R. *Personal communication*, 1982.

5 LEVI, L., ed. *Society, stress and disease: working life.* Oxford, New York, and Toronto, Oxford University Press, 1981, Vol. 4.

6 LEVI, L. ET AL. Work stress related to social structures and processes. In: Elliott, G. R. & Eisdorfer, C., ed. *Stress and human health: analysis of implications of research.* New York, Springer, 1982, pp. 119–146.

7 United States of America. President's Commission on Mental Health. *Report to the President.* Washington, DC, United States Government Printing Office, 1978, Vol. 1.

www.ingramcontent.com/pod-product-compliance
Lightning Source LLC
Chambersburg PA
CBHW081808200326
41597CB00023B/4185